Memoir of Localities of Minerals of Economic Importance and Metalliferous Mines in Ireland.

BY

GRENVILLE A. J. COLE, F.R.S., M.R.I.A.

A facsimile edition of the Geological Survey of Ireland Memoir of 1922.

The Mining Heritage Society of Ireland
Dublin: 1998

© This third edition: Mining Heritage Society of Ireland
© First 1922, and second 1956 edition: Geological Survey of Ireland

This edition is published by the Mining Heritage Society of Ireland 1998,
c/o Geological Survey of Ireland,
Beggars Bush,
Haddington Road,
Dublin 4,
Ireland.

Reproduced and printed by ColourBooks Ltd., Baldoyle, Dublin 13
from an original in the Department of Geology, Trinity College, Dublin
(used with the kind permission of Dr Patrick Wyse Jackson).

Front cover: The Man-Engine House, Mountain Mine, Allihies, Co. Cork. Cole, pp. 46 - 47. This is the only Man-Engine House ever constructed in Ireland, apparently by Michael Loam, the designer of Cornish Man-Engines. It was erected, as a combined winding and Man-Engine house, in 1862 and was powered by either a 32" or 36" cylinder diameter Cornish engine to an operational depth of 1,494 feet. The mine was closed in 1882 and although the engine house has lain derelict ever since, it is still remarkably intact, retaining substantial portions of the engine house, chimney and, most unusually, the boiler house. The engine house is, however, in imminent danger of collapse due to the removal of masonry and window lintels near one corner. It is estimated that less than 25 Man-Engines were ever constructed out of a world wide total of some 3,000 Cornish engine houses of all types (N. Johnson, County Archaeologist, Cornwall, pers. comm.). Consequently, not only is this an extremely rare example of Cornish mining technology, it is arguably one of the most intact Man-Engine Houses still left in existence anywhere in the world, and, therefore of major national and international heritage value. (Photograph: Charles Tyrrell, 1998).

Rear cover: Tramway Arch, West Avoca, Co. Wicklow. Cole, pp. 35, 140. This bridge forms part of the steam operated tramway constructed by Henry Hodgson in 1846 to convey pyritic ore from his mines in West Avoca (Wicklow Copper Mining Co.) to the port in Arklow. This was one of the earliest steam operated railways in Ireland, and was finally taken over by the Dublin, Wicklow and Wexford Railway in May 1861. The chimney of the Ballygahan Engine House is visible through the archway, and in the far distance, the slopes of East Avoca. Although the Tramway Engine House chimney is still standing, sadly the remains of the associated Engine House were apparently demolished as recently as 1996. (Photograph: Matthew Parkes, 1997).

CONTENTS

Foreword	Matthew Parkes	v
Memoir of Grenville A.J. Cole	Patrick Wyse Jackson	vii
Captions to Figures 1-23		xvii
Figures 1-23		xxiii
Memoir of Localities of Minerals of Economic importance and Metalliferous Mines in Ireland.	Grenville A.J. Cole	1

LIST OF FIGURES

Figure 1. Williams Engine House, Avoca, Co. Wicklow	xxiii
Figure 2. Hollyford Mine, Co. Tipperary	xxiv
Figure 3. Tankardstown, Bunmahon, Co. Waterford	xxiv
Figure 4. Mountain Mine, Allihies, Co. Cork	xxv
Figure 5. "Cappagh" Copper Mines, Co. Cork	xxvi
Figure 6. Crookhaven Mines, Co. Cork	xxvii
Figure 7. Glandore Mines, Co. Cork	xxviii
Figure 8. Creevelea, Co. Leitrim	xxviii
Figure 9. Glendalough, Co. Wicklow	xxix
Figure 10. Conlig, Co. Down	xxix
Figure 11. Keeldrum Mine, Co. Donegal	xxx
Figure 12. Conlig/Whitespots, Co. Down	xxxi
Figure 13. Lisdrumgormly, Castleblayney, Co. Monaghan	xxxii
Figure 14. Glengowla Mine, Oughterard, Co. Galway	xxxiii
Figure 15. Hope Mine, Castleblayney, Co. Monaghan	xxxiii
Figure 16. Ballycorus Mine and Leadworks, Co. Dublin	xxxiv
Figure 17. Hero Mine, Luganure, Co. Wicklow	xxxv
Figure 18. Connemara Mining Company of Ireland	xxxvi
Figure 19. Ballyhickey Lead Mine, Quin, Co. Clare	xxxvii
Figure 20. Kilbricken Mine, Quin, Co. Clare	xxxvii
Figure 21. Shallee, Co. Tipperary	xxxviii
Figure 22. Barrystown, Bannow, Co. Wexford	xxxviii
Figure 23. Dhurode Mine, Co. Cork	xxxix

FOREWORD

This facsimile of a 1922 memoir on the mineral localities and metalliferous mines in Ireland comes at a time when a great upsurge of interest in local heritage means it will assist many people with their own projects. Despite a substantial output of new maps and publications from the Geological Survey of Ireland, this Memoir is still one of the most requested items by members of the general public. Although reprinted in the 1950s it has remained out of print for many years and it is often difficult for people to access. The reprint has come about through collaboration of several parties, all of whom saw the need to make it available again.

This edition is a straight facsimile of the original text, but we have left out the Map that accompanied the original in order to keep the price of this edition reasonably low. It was large, colour printed, but with bland shades, and would have been expensive to reprint and to bind. Topographic maps are now widely available, especially the new Ordnance Survey 1:50,000 maps which show townlands, such as those mentioned in the text. Any more detailed research would probably require the use of older 6 inch maps in any case, or the original Geological Survey of Ireland fieldsheets which can be consulted in the Public Office of the GSI.

The original publication is supplemented by a short biography of Grenville Cole, the author of the Memoir, who was a major figure in Irish Geology in the late 19th and early 20th centuries. This reprint is a tribute to the continuing relevance of his work to mine heritage, as well as to geological studies. A selection of mine plans and sections that illustrate some 19th century mines in Ireland from the Historic Mine Records in the Geological Survey of Ireland are incorporated as examples of records available for further research. The Director of the Geological Survey, Dr Peadar McArdle has generously agreed to waive the Survey's copyright for this reprint, and we gratefully acknowledge the ongoing support of the Geological Survey of Ireland towards the work of the Mining Heritage Society of Ireland.

The Society is currently conducting an inventory of all extant mining heritage sites in Ireland, and this book records many of the smaller sites, too easily forgotten without visible engine houses or open workings. In the future, we anticipate that new publications of current information may be possible as a result. This book provides a starting point for these inventory efforts. Outside of specialised mining heritage interests, those who will hopefully find it most useful are groups concerned with local history and the heritage of their own area.

As an example of its potential use, I recount the following tale. I was asked by the Secretary of the Valentia Heritage Society, Mrs Ring, to identify the source of information for a comment on copper mining in Valentia Island in a publication on the geology of Kerry, by Patrick Wyse Jackson of the Geology Museum in Trinity College Dublin. I immediately thought of Cole's *Memoir*, and indeed it turned out to be so. This provided the information that was sought as to where the copper mining had taken place on the island, and when ore was extracted, the scale of working and so on. Enough information was contained in that short entry to link the copper mining to some local folklore. Reputedly, potatoes in that particular locality escaped the potato blight during the famine, possibly due to copper in the local soil.

The Mining Heritage Society of Ireland have produced this book to assist community groups in researching and preserving their own local heritage. The Society has assisted and supported several existing groups in this work in minor ways, but we are open to new members, new ideas and new requests for help in whatever way we can, to fulfil the aims of the Society in developing awareness and appreciation of Ireland's rich mining heritage.

This publication has received support from the Heritage Council under the 1998 Publications Grant Scheme. We welcome the Council's recognition of this important aspect of industrial and social heritage which has shaped both the landscape and its communities in many parts of Ireland.

We are grateful to Rob Goodbody, William Dick, Norman Moles, Ken Brown, Charles Tyrrell, Martin Critchley, John Morris, Des Cowman, Rita Kirkpatrick and Gerry Stanley for their invaluable assistance with photographs and illustrations.

Matthew Parkes,
Secretary, Mining Heritage Society of Ireland, August 1998.

GRENVILLE ARTHUR JAMES COLE (1859-1924): A BIOGRAPHICAL MEMOIR

Patrick N. Wyse Jackson

Grenville Arthur James Cole (1859-1924) was one of the foremost geologists working in Ireland at the turn of the twentieth century. Although of English extraction he spent the majority of his working life in Ireland, where he was Professor of Geology and Mineralogy at the Royal College of Science in Dublin, as well as Director of the Geological Survey of Ireland. He also became deeply involved in a number of scientific societies in Dublin.

His work on igneous and metamorphic geology, Irish regional and off-shore geology, and geographical studies were probably his most important. However he never produced any work of major scholarship, although he wrote much that enhanced the understanding of his subject, and was known, through his popular textbooks, to a generation of geology graduates. He received two awards from the Geological Society of London, was conferred with the degree of Doctor of Science from Queen's University Belfast, and was elected to Fellowship of the Royal Society.

While he was popular with many contemporaries he did not endear himself to some. This was due to his ambitious nature and his occasional insensitivity to the feelings of other scientists.

He was a man of endless energy, a widely travelled tricyclist and later, bicyclist who spoke both French and German, a fine field geologist who could write readable texts, a prolific author, a neat draftsman, and a noted lecturer and teacher.

GEOLOGICAL APPRENTICESHIP

Cole was born in London on the 21st October 1859, and educated at the City of London School, and at the Royal School of Mines. In October 1878 he was appointed as Demonstrator to the School, where his duties included planning and teaching practical classes, and carrying out research on igneous and metamorphic rock textures. From 1886 to 1890 he also was Head of Geology at Bedford College for Women (since incorporated into the University of London) - subsequently a student geological society was named in his honour.

Around this time Cole began to travel, at home, and on the

Continent, and the majority of these trips were made by tricycle and later by bicycle. In his entry in the 1924 edition of *Who's Who* he stated his only pastime to be "cycling especially as a means of travel". In 1894 he published *The Gypsy Road,* a delightful account of a trip to Poland.

DUBLIN ACADEMIC

In mid-1890, at the age of thirty, Cole was appointed to the Chair of Geology at the Royal College of Science in Dublin, and was later to take up positions at the Museum of Science and Art as Curator of Geology, and in the Geological Survey of Ireland as Director. From 1905 he held each of these posts concurrently. He was also active in various Dublin scientific societies including the Royal Dublin Society, the Royal Irish Academy, the Dublin Naturalists' Field Club, the Dublin Microscopical Club, and the Irish Geographical Association.

Cole was drawn into the company of a scientific group that included William Johnson Sollas (1849-1936) the geologist, Alfred Cort Haddon (1855-1940) the zoologist and anthropologist, and Robert Lloyd Praeger (1865-1953) the botanist. This scientific circle used to meet daily in the Misses Gardiner's *Farm Produce Depot and Refreshment Rooms* at 1-2 South Leinster Street for discussion, tea and buttered toast - how many important scientific theories were formulated there one can only speculate. On Cole's initiative, their hostesses were presented with seven marble-topped tables on which to serve their fare. These were described by Cole in an article entitled *Geologists at the Luncheon-Table* published in *The Irish Naturalist* in 1895.

COLE'S SCIENTIFIC WORK
The Royal College of Science

Initially Cole found the number of geological classes and the number of students to be small, with often only one or two of the latter in the third year. Annually he delivered 80 lectures to various groups: agricultural students, science students and student-teachers, and soon after his arrival in Dublin he increased the number of practical classes and field trips. The annual excursion for 2nd year agricultural students (or "scholars" as he called them) was by no means an easy venture: the party would depart on the early train at 5 or 6 o'clock in the morning to Sligo, Waterford, or Galway, after which they would cycle up to twenty five miles, frequently stopping to examine some outcrops, before retiring for the night!

In 1908 Cole and George Herbert Pethybridge (1871-1948), a botanist at the college, ran a summer course in Rural Economy for teachers from technical and national schools. In 1919 Cole taught two courses in general science for men recently demobilised from the army.

In the early years, conditions at the Royal College of Science then located at 51 St Stephen's Green were very cramped and the geological laboratory was too small to be truly effective. This problem was solved when new College buildings on Upper Merrion Street were opened by King George V in 1911. In 1913 his salary was £800 per annum.

The Museum of Science and Art and geology in schools

As Custodian of Geology in the Museum of Science and Art (1896-1924) Cole had access to many important collections. In his eagerness to encourage the teaching of geology in some secondary schools he made loans of material from the museum. In 1952 a number of specimens were returned to the National Museum of Ireland (formerly the Museum of Science and Art) which transpired to be part of the important Leskean Collection!

Cole did not spend as much time working on the collections as he should have done, and fell foul of some Museum staff members, particularly Robert Francis Schraff (1858-1934), who was Custodian of the Palaeontological Collections. In 1906 Cole was given an attendance book, and was expected to spend two days a week working on the collections. Soon afterwards Cole was left to tend only the collection of Irish minerals.

Geological Survey of Ireland

In 1890 the Irish Geological Survey published the last of the one-inch-to-one mile coloured geological maps and its activities were run down. Routine work such as public inquiries constituted most of the work for the five remaining geologists. Sollas, and later Cole were employed on a part-time basis as petrologists, on a renumeration of 30 shillings a day.

On the 1st April 1905 the Geological Survey ceased to be a divisional branch of the Geological Survey of Great Britain and Ireland and became incorporated into the Department of Agriculture and Technical Instruction for Ireland. Cole, the only geologist in Ireland of international standing other than John Joly, was appointed part-time Director. In 1920 he earned a salary of £100 per annum while the District Geologist earned £600 and the Geologists £120 per annum. Towards the end of the nineteenth century there was a flurry of oceanographic research beginning with the *Challenger* expeditions of 1872-76. Zoologists were busily dredging Irish waters in search of elusive species, and geologists were soon to follow in search of rock. The Department of Agriculture and Technical Instruction's protection vessel the *Helga* trawled the Porcupine Basin for rock samples. The *Helga* was later used as a gun-boat and shelled O'Connell Street from the River Liffey during the Easter Rising of 1916. During these 'troubles' all the back issues of *The Irish Naturalist*, a magazine to which Cole frequently contributed, were destroyed when the premises of Easons were hit. The Porcupine rock dredgings were described by Cole and Thomas Crook (1876-1937) in a Department of Fisheries report of 1903 and again in a Geological Survey publication of 1910. This publication is important because it raised the possibility of Mesozoic and Tertiary rocks offshore of Ireland. Over half a century was to follow before Ireland's offshore regions were next surveyed during the search for oil. Drilling during this phase of exploration confirmed the presence of post-Palaeozoic rocks in offshore basins.

Mapping was not neglected however during Cole's Directorship of the Geological Survey. Thirty eight one-inch-to-one-mile maps affected by the then Director Edward Hull's revision of the geology of Munster (1878-81) were re-edited and published in 1913. This work was necessary because Hull had interpreted the Dingle Beds as upper Silurian in age, and had rejected the earlier Devonian age determinations. While Cole was Director four new smaller scale maps at 1/4-inch-to-one-mile of Lough Foyle, Belfast-Lough Neagh, Dublin, and Cork, and a larger scale six-inches-to-one-mile geological and drift map of Dublin all appeared. There was to have been a complete coverage of the country at the scale of 1/4-inch-to-the-mile but the plans were abandoned in 1922. Publication of the memoir that accompanied the revised Killarney and Kenmare map was delayed and it did not appear until 1927.

The Survey's geologists examined the intrabasaltic rocks of northeast Ireland in search of commercial deposits of bauxite in 1907 and 1908. A fuel map showing the extent of coal fields and peat bogs (1921), a map and memoir outlining the minerals and mines (1922) and memoirs on Barytes (1923) and the Ballycastle Coalfield (1924) were published by the Geological Survey under Cole's Directorship.

Royal Irish Academy

Cole was elected a member of the Royal Irish Academy on the 13th of February 1893. He was a member of the Council for a number of years and was Librarian between 1899 and 1905. Many of his important papers were published in the *Proceedings* and *Transactions* of the Academy including those on metamorphic rocks of Tyrone and Donegal (1900) and gneisses of Donegal (1902). He played an important role in the Clare Island Survey (1909-1911) as a member of the organising committee. The paper on the geology of the island was written by Hallissy. Cole helped produce a geological map and memoir of the island, which was published by the Geological Survey in 1914.

Royal Dublin Society

Cole also published in Royal Dublin Society journals, on basaltic andesites (1894), rhyolites (1896), the geology of Slieve Gullion (1897), and orbicular granite of Donegal (1916). He also gave many public lectures at the Society.

Dublin Naturalists' Field Club

Cole had a long and successful association with the Dublin Naturalists' Field Club which he joined in 1890. Soon he was lecturing and leading trips for the Club; to Portrane, the Scalp and Tibradden, and later to Dundalk, Fermoy and Antrim. Much of his knowledge of the geology of Ireland was gained while on these trips and many of his papers of this time reflect this influence - *County Dublin, Past and Present* (1892), *On Variolite and Other Tachylytes at Dunmore Head, Co. Down* (1894), and *Scenery and Geology in County Antrim* (1895). He gave many talks and a popular feature of these was his use of excellent lantern slides. He was able to bring his experience and knowledge into the lecture theatre and was, by all accounts, an excellent lecturer. He was elected President in 1896-97.

In a lecture in late 1897 he asserted that some form of geological conservation was necessary and he suggested that the Giant's

Causeway be closed off and made accessible only to the paying public! This classic geological location has recently been designated a UNESCO World Heritage Site. Cole was one of the first people to show concern for conservation.

Dublin Microscopical Club

This was a small club which was founded in 1849 for social and microscopic purposes. In the 1890s when Cole was invited to join the club it consisted of twelve members and so could meet in the house of one member every month. Cole usually exhibited rocks that contained interesting petrological features. These contributions are described in the pages of *The Irish Naturalist*. The Club had ceased to function by 1924.

Irish Geographical Association

Through his observations while on field excursions and bicycling trips Cole realised the importance of the interaction between geology and geography. After 1910 many of his papers deal with the description and development of geomorphological features. He was a founding member in 1918 of the Irish Geographical Association of which he was President from 1919 to 1922. He was also President of the Geographical Association (of Great Britain) in 1919. Cole played an important part in the advancement of geography in the British Isles, which by the beginnings of the 1900s was generally adopted as a school and university subject.

MARRIAGE AND FAMILY LIFE

In 1896 Grenville Cole married Blanche Vernon, of Clontarf Castle, and moved to 3 Uxbridge Terrace (now 3 Dartmouth Square). She was to become a constant bicycling companion and co-author of *As we ride (1902)*, an account of a Balkan trip. They had one son, Vernon. He later joined the East African Civil Service, but returned to Dublin to practise as a medical doctor. In 1902 Cole purchased two houses, 10 Winton Road, Leeson Park, and *Orohova* (now *Glenheather*) in Carrickmines. A man of great stamina, despite standing just over five feet tall, he regularly walked from Carrickmines to Dublin, a distance of eight miles, in the morning, returning in the evening by foot. This is quite surprising when you consider that Carrickmines railway station, on the ill-fated Harcourt

Street line, was a short distance across the fields from his house.

ILLNESS AND HONOURS

Towards the end of his life rheumatoid arthritis progressively crippled Cole, but through considerable pain and disability he continued working. In 1909 he received the Murchison Medal of the Geological Society of London. He was President of Section C (Geology) at the 1915 British Association for the Advancement of Science. Queen's University, Belfast conferred upon him the degree of Doctor of Science (his only degree), and in 1917 he was elected a Fellow of the Royal Society. He died at his home in Carrickmines on 20th April 1924, and is buried in Deans Grange Cemetery. His wife died on 17th November 1927 and is buried beside him.

PUBLICATIONS

Cole was a very prolific author in a wide diversity of geological topics: petrology, stratigraphy, palaeontology, geomorphology, and structural geology. A full bibliography is given in Wyse Jackson (1989).

TABLE 1 Cole's publications.

Subject	Number of papers	Subject	Number of papers
Reviews	150	Obituaries	11
Regional geology 77 (Irish 53, Others 24)		Dredging and geology	7
Igneous and metamorphic studies	74	Sedimentary geology	7
Geomorphology and geography	40	Education & Lecturing	6
Petrology and mineralogy	29	Geological mapping & GSI	5
Palaeontology	27	Exploration of Irish Caves	4
General interest	22	Meteorites	3
British Association Reports	20	Collections Museum Science Art	2
Structure Earth & Earth Movements	16	Geological Conservation	1
Economic aspects	13	Poetry	1
Books	12	**TOTAL:**	**527**

His earliest research papers dealt with volcanic rocks and petrological textures. In his first paper Cole demonstrated the production of perlitic structure, which is often seen in basalts, by quenching molten Canada Balsam with water. During his first years in Ireland he continued to research into igneous and petrological problems. He published a paper on the occurrence of Riebeckite in Britain, in which he distinguished between that carried southwards from Ailsa Craig (by ice) and that occurring *in situ*.

Cole's interest in field studies is reflected in the large number of regional geological accounts that he produced, and by his textbook *Open-Air-Studies: an introduction to Geology out-of-Doors*. The majority of these regional accounts were Irish but some resulted from his cycling trips on the Continent such as papers on the geology of the Austria-Hungary Empire.

He did much research on the Dalradian metamorphic belt in Donegal, and was also interested in granites. He described orbicular granite in Donegal, and intrusions in Tyrone and Londonderry. In an excellent paper he outlined the structure of Slieve Gullion and also focused his attention on the rhyolites of County Antrim and on other volcanic rocks of Ulster.

In 1892, just two years after his arrival in Ireland, he penned a series of illustrated articles on the geology and scenery of Dublin, that appeared in *The Irish Naturalist*. He was then commissioned, in 1895, by the owners of the North Eastern Railway, to write a similar booklet outlining the geology and scenery as seen from the railway carriages. It was the first booklet of its kind. In it Cole suggested that the traveller should examine the photographs of Robert John Welch (1859-1936), the Belfast professional photographer who had an interest in molluscs and who was very active in the Belfast Naturalists' Field Club.

A feature of Cole's later writings is the attention given to geographical matters. He described many geographical and social features in two travel books: *The Gypsy Road* and *As We Ride*. He examined coastal features of Ireland for a Royal Commission on Coastal Erosion in 1911. He produced two fine books outlining geological influence on man and the landscape: *The Growth of Europe* and *Ireland, the Land and the Landscape*. Both appeared in 1914.

In three general papers he described the glaciation of Ireland and indicated that glacial deposits at Killiney had undergone subaqueous reworking since deposition. He accepted the subglacial origin of eskers and in a lecture delivered in 1892 he discussed the origin of the Scalp in County Dublin. He reiterated Hull's explanation of its formation: namely that it was excavated by a southward flowing stream during the Carboniferous. By 1911 he had accepted

G.W. Lamplugh's explanation: that the Scalp was a glacial lake spillway. In 1912 he wrote about the Liffey Valley and concluded that the flow of the River Liffey and the King's River must have undergone a post-glacial reversal. The Slade at Saggart and the misfit valley of the Brittas River marked the former course of the Liffey. Perhaps his most important paper on glacial features appeared after he had travelled to Spitsbergen in 1911. He observed active glaciers with the resultant deposits and attempted to explain the mechanics of deposition for Irish glacial material. He erred when he decided that the glaciation of Killary Harbour was much earlier than that which affected Killarney. This was based on the observation that the corries of Killarney are less eroded; he did not take into account different rock types.

He made a number of incursions into palaeontological research. In one short publication he defended the Carboniferous age of a trilobite from Kiltorcan, County Kilkenny. In his most important palaeontological paper paper he used his skill as a petrologist to elucidate the structure of the bryozoan *Hemitrypa hibernica*. From thin sections he accurately described the structural relationship between the reticulate colony and its protective superstructure, which had been described as being parasitic by earlier workers.

He was interested in educational and curricular matters. He called for the introduction of geology to the secondary school syllabus declaring it to be more useful than classics. He wished to see agricultural students getting a grounding in soil science, and attempted to encourage the teaching of geography in the class room and in the field.

During 1907 and 1908 field studies carried out by the Geological Survey of Ireland evaluated the extent and economic potential of these deposits. The first results of this work were presented to the Dublin meeting of the British Association Meeting in 1908. However the publication of all the results did not appear until 1912. Although there were a number of small active mines, the deposits were too limited to support large-scale extraction.

In 1916, in response to War Office worries about resources, Cole gave notice of the intention to produce a Memoir that would review the deposits of economic value. The Memoir (reproduced herein) was published in 1922 and is accompanied by a large annotated map. It was the first serious account of Irish deposits since George Henry Kinahan's (1889) *Economic Geology of Ireland*.

In 1901 the Porcupine Bank was dredged and the derived samples were subsequently described. This was one of the first publications of off-shore geological studies. Although many of the specimens examined were granite and gabbro, flint and chalk pebbles were also present. Further dredging in August 1907 off County Kerry yielded rock that suggested the presence of Cretaceous deposits.

Possibly the most appealing aspect of Cole's work was his literary style which makes for easy relaxing reading. He wrote poetry, some of which was published in 1911. These poems describe the history of a Templar hospice in France.

SUGGESTED FURTHER READING

[HERRIES] DAVIES, G.L. 1977. The Making of Irish Geography, II: Grenville Arthur James Cole (1859-1924). *Irish Geography* **10**, 90-94.

HERRIES DAVIES, G.L. 1983. *Sheets of Many Colours.* Royal Dublin Society.

HERRIES DAVIES, G.L. 1995. *North from the Hook.* Geological Survey of Ireland.

PRAEGER, R.L. 1941. *A populous solitude.* Hodges Figgis, Dublin.

WYSE JACKSON, P.N. 1989. On Rocks and Bicycles: a biobibliography of Grenville Arthur James Cole (1859-1924) fifth Director of the Geological Survey of Ireland. *Geological Survey of Ireland Bulletin* **4**, 151-163.

WYSE JACKSON, P.N. 1991. The cycling geologist. *Cycle Touring and Campaigning* **June-July 1991**, 26-27.

This memoir is an abridged version of my paper of 1989 to which a new section on Cole's publications has been added but with the bibliography omitted.

CAPTIONS TO FIGURES 1-23

Figure 1. Williams Engine House, Tigroney, Avoca, Co. Wicklow. [Cole, pp. 34, 84, 140]
This building is arguably the finest example of Cornish Engine House architecture still standing in Ireland. It was constructed by the Williams family, owners of the Perran Foundry, Cornwall, in 1850 and housed a 60" engine which was finally removed from the building in 1884. The building, and the very fine chimney, still remain in remarkably good condition and appear very little different to their appearance in 1956, apart from the growth of surrounding trees. Actions to consolidate and conserve this building were initiated in 1997. (Photograph: J.C. Ferguson, 1956, from the collection of Kenneth Brown, Hayle, Cornwall)

Figure 2. Hollyford Mine, Co.Tipperary. [Cole, p.39]
The chimney is now all that remains at this former copper mine. (Photograph: Des Cowman, 1997)

Figure 3. Tankardstown, Bunmahon, Co. Waterford. [Cole, pp. 42-43]
The ruins of two engine houses and part of a single chimney and boiler house dominate the cliffscape east of Bunmahon. The larger and more intact ruin is believed to be that of a pumping engine house, the other that of a winding engine. Both engines were powered from a single boiler and, equally unusually, the pumping engine house is plastered on the exterior. This has undoubtedly contributed to the survival of the ruin, which was constructed around 1850. The building has 'starred' in two film productions, in the 1960s, when it featured in the film *The Mackenzie Break* and again in the 1970s in a Swedish film production. For the latter it was transformed into Count Dracula's Castle and in the former it featured, towards the end of the film, as the site for the disposal of a burning lorry down the pumping engine shaft. The lorry is still there, albeit covered by back fill to surface.(Photograph: Martin Critchley, 1997).

Figure 4. Mountain Mine, Allihies, Co. Cork [Cole, p. 46-47]
Part of a hand water coloured cross section of Mountain Mine, Allihies, Co. Cork, from 1878. The Man Engine house is shown on the right (see front cover picture and caption). The engine house on the left has collapsed into the shaft in recent years. [GSI Mine

Records Box MR27, Item 5]

Figure 5. "Cappagh" Copper Mines, Co. Cork. [Cole, p. 53]
A small part of a hand water coloured section of "Cappagh" Copper Mines in Co. Cork, from approximately 1857, drawn and surveyed by John [XXX], F.G.S. Civil and Mining Engineer. [GSI Mine Records Box MR38, Item 2]

Figure 6. Crookhaven Mines, Co. Cork. [Cole, p. 54]
Showing most of a beautiful hand water coloured cross section through Crookhaven Mines in Co. Cork, dating from 1863. The following note is recorded to the side of the portion illustrated:- "The whole of the lodes and so called Elvan are interstratified with the sedimentary beds which are coarse and of an unproductive nature, hard and almost crystalline grits. some carbonate of copper has been found on the back of Purple lode with a few spots of purple sulphuret." [GSI Mine Records Box MR40, Item 3]

Figure 7. Glandore Mines, Co. Cork. [Cole, p. 60, 88, 130]
An engine house at these mines where copper, manganese and iron have been worked. (Photograph: Martin Critchley, 1996).

Figure 8. Creevelea, Co. Leitrim. [Cole, pp. 77-78]
Iron ore was first smelted at Creevelea in the mid-1600s and the works extended in the 1690s. Smelting trials using local coal were conducted in the late 1760s and with coking coal in the 1830s. By 1866, the plant consisted of two hot-blast smelting furnaces, processing iron ore derived from Tulnamoil and Gowlan townlands, and operated at considerable profit. A very fine photograph, dated 1905, in the possession of the current site owner, shows a very extensive, and intact, iron works infrastructure. Sadly, the furnace illustrated, and an engineer's house, is all that now remains, the rest of the works having being demolished in relatively recent times to provide a source of road construction material. (Photograph: John Morris, 1998)

Figure 9. Glendalough, Co. Wicklow. [Cole, p. 84, 110, 112]
Different generations of mining leave their imprint on mining landscapes. William Dick and his dog are here standing beside a Cornish rolls crusher, made in Wales, from a 1950s phase of

working. (Photograph: Martin Critchley, 1997).

Figure 10. Conlig, Co. Down. [Cole, p. 94]
10a. View looking northwest of the South Engine Shaft chimney stack and, in the foreground, one of three tailings empoundments? downslope from the dressing floors. The tailings are currently subject to intense erosion due to the activities of motor bike scramblers.
10b. View looking southeast of the Bog Shaft at Whitespots. Sunk to improve drainage at the south end of the mineral lode, the shaft was probably not used to lift ore, as the spoil heap is barren. The engine house has partially collapsed [see Figure 12]. (Photographs: Norman Moles, 1998)

Figure 11. Keeldrum Mine, Co. Donegal. [Cole, p. 91]
Part of a section on Jacob's Lode, from a section and plan of Keeldrum Mine, Co. Donegal, date unknown. [GSI Historic Mine Records Do 01 111]

Figure 12. Conlig/Whitespots, Co. Down. [Cole, p. 94]
12a. Lead Mines in 1948, with McNulty's House on the right. The windmill stump at Whitespots, was built as a flour mill in 1780 and subsequently converted to power a stamp mill.
12b The engine house and unusual square chimney at Bog Shaft in 1948.
12c. An engine house at Conlig in 1949, no longer present. This was probably at the North Engine Shaft. (Photographs Noel Kirkpatrick, 1948/9)

Figure 13. Lisdrumgormly, Castleblayney, Co. Monaghan. [Cole, p. 99].
This relatively small lead mine was apparently first worked some time prior to 1836 and subsequently re-examined by the Farney Development Co. in the 1920s and again by the Mining Corporation of Ireland in 1956. The plan illustrated, prepared by M.V. OBrien, then Director of the Geological Survey of Ireland, is dated 1956. and provides a wealth of detail of the underground extent of this mine. The inset photograph shows the site in 1984: the buildings, dating from the 18th Century, were apparently used by the Mining Corporation as a core store, and the mine shaft was located in the

field immediately to the right of the vehicle. [Plan: GSI Historic Mine Records] (Photograph: John Morris, 1984)

Figure 14. Glengowla Mine, Oughterard, Co. Galway. [Cole, p. 122, 143].
The Geoghegan family have recently restored the workings and buildings of this 1850s silver and lead mine and opened it to the public as a Show Mine, with underground tours. View shows reconstructed timber horse whim to left; timber windlass over the whim shaft; office building and blacksmiths shop to rear.

Figure 15. Hope Mine, Castleblayney, Co. Monaghan. [Cole, pp. 100 - 101].
This mine apparently operated between 1852-1869, and perhaps as early as 1849, although records are somewhat confusing as the workings are listed under various names and locations. Very little now remains to be seen, apart from the engine house chimney and some building and yard wall foundations only a few courses high. However, an adit leading to an internal winze, flooded to adit floor level, was discovered during land drainage work in the early 1980s. (Photograph: John Morris, 1984)

Figure 16. Ballycorus Mine and Leadworks, Co. Dublin. [Cole, p. 107]
16a. The entrance to the leadworks taken around 1905. The little boy on the right was the son of the manager, named Roberts.
16b. The Shot Tower at Ballycorus in 1905. Molten lead was poured, from the top of the tower, through a sieve to create droplets, which were quenched in a water pond at the base.
16c. The Ballycorus Chimney in 1905 with brickcourse at the top, and the damper, which is at the end of the very long flue from the works. (Photographs: by an unknown photographer, from the collection of William Dick)

Figure 17. Hero Mine, Luganure, Co. Wicklow. [Cole p. 110]
17a. A view looking eastwards over the main area of mining remains at the Hero Mine, Luganure, in Glendasan, Co. Wicklow. The remains can be seen of the crushing mill, dressing floors where lead ore was broken and picked out by hand, and the spoil heaps of the Fox-rock Mine in the distance across the valley. (Photograph:

Matthew Parkes, 1997).

17b. Middle: Close up of a buddle, at Hero Mine, Luganure, Co. Wicklow. This was a method of water powered, rotating, gravity separation of finely crushed ore from gangue, or waste mineral. The central, polished wooden spindle is still visible. (Photograph: Matthew Parkes, 1997)

17c. A detail picture of the crushing mill and a cobbled dressing floor where lead ore was broken and picked out by hand. (Photograph: Rob Goodbody, 1994)

Figure 18. Connemara Mining Company of Ireland. [Cole. p.116]

This company, capitalised at £15,000 in £1 shares, is most directly linked with the development of the Caherglassaun (Caherglissane) silver-lead deposit, near Gort in Co. Galway. The deposit was discovered in early 1850 and rather grandiosely compared, by the *Galway Vindicator*, with the richest deposits of S. America. By 1852 the deposit is associated with those at Glengowla [Cole, p.102, 143] and the Curraghduff and Glann Mine [Cole, p. 29] in a report by Pierre J. Foley (*Mining Journal*). Substantial underground developments are reported, along with the erection of an engine, and by 1854, developments to the 25 fathom, "Tennent's Venture", level are noted. This name is almost certainly that of R.D. Tennent, the first signatory director on the share certificate. The demise of the Company appears to be marked by a brief report of litigation against four shareholders (the signatory directors and Secretary?) in the *Mining Journal* of November, 1855. (Original Certificate, for 5 One Pound shares, dated July 26th, 1852, collection of John Morris).

Figure 19. Ballyhickey Lead Mine, Quin, Co. Clare. [Cole, p. 119]

The remarkably well preserved, very slender and graceful engine house chimney dates from the 1834-1838 period of mining a rich pocket of lead - silver ore by open pit. (Photograph: John Morris, 1998).

Figure 20. Kilbricken Mine, Quin, Co. Clare. [Cole, pp. 15, 119-120]

The Kilbricken lead - silver deposit was discovered in 1833 and operated, under various owners, including John Taylor and Co., the Kilbricken Silver and Lead Co., and the Clare United Silver and Lead Co., before final closure in 1856. Some of the infrastructure developed is still readily visible, including a cobbled dressing floor, magazine, various mine buildings, and part of one of three pumping engine houses. The photograph shows all that remains of that engine house, probably constructed in the mid-1840s to house a 50" engine. Clare Calcite Ltd currently (1998) propose to redevelop the site as a calcite mine. (Photograph: Martin Critchley, 1998)

Figure 21. Shallee, Co. Tipperary. [Cole, p. 122]
The 1860s engine house for pumping and for crushing power, was re-used as an ore bin in the 1950s, whilst the left front building was used for processing plant. (Photograph: Martin Critchley, 1990)

Figure 22. Barrystown, Bannow, Co. Wexford. [Cole, p.125]
This lead mine may have commenced in the 8th century, but little now remains except this engine house. (Photograph: Martin Critchley, 1997)

Figure 23. Dhurode Mine, Co. Cork. [Cole, p.55]
23a. The magazine or powder house with workings behind. (Photograph: Matthew Parkes, 1997)
23b. Stone steps built by miners to get to a small working in the cliff face. (Photograph: Matthew Parkes, 1997)

Figure 1. Williams Engine House, Avoca, Co. Wicklow

Figure 2. Hollyford Mine, Co.Tipperary

Figure 3. Tankardstown, Bunmahon, Co. Waterford

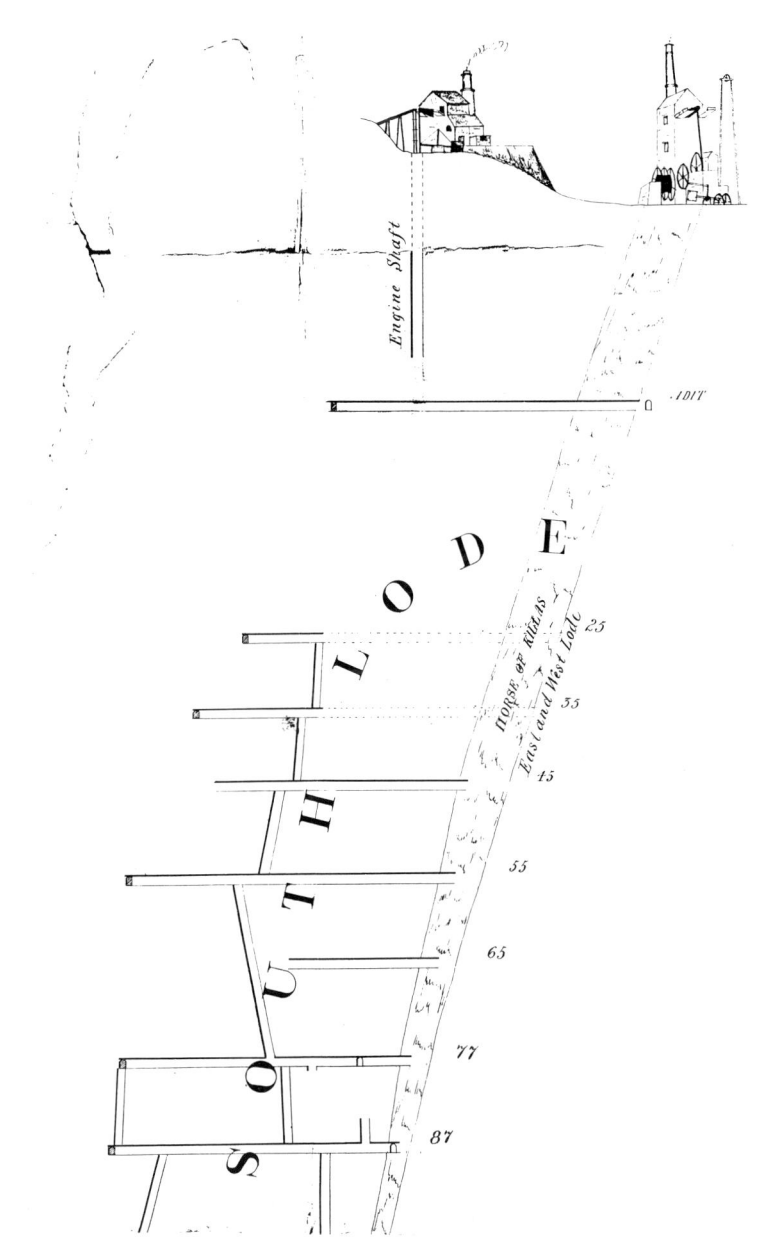

Figure 4. Mountain Mine, Allihies, Co. Cork

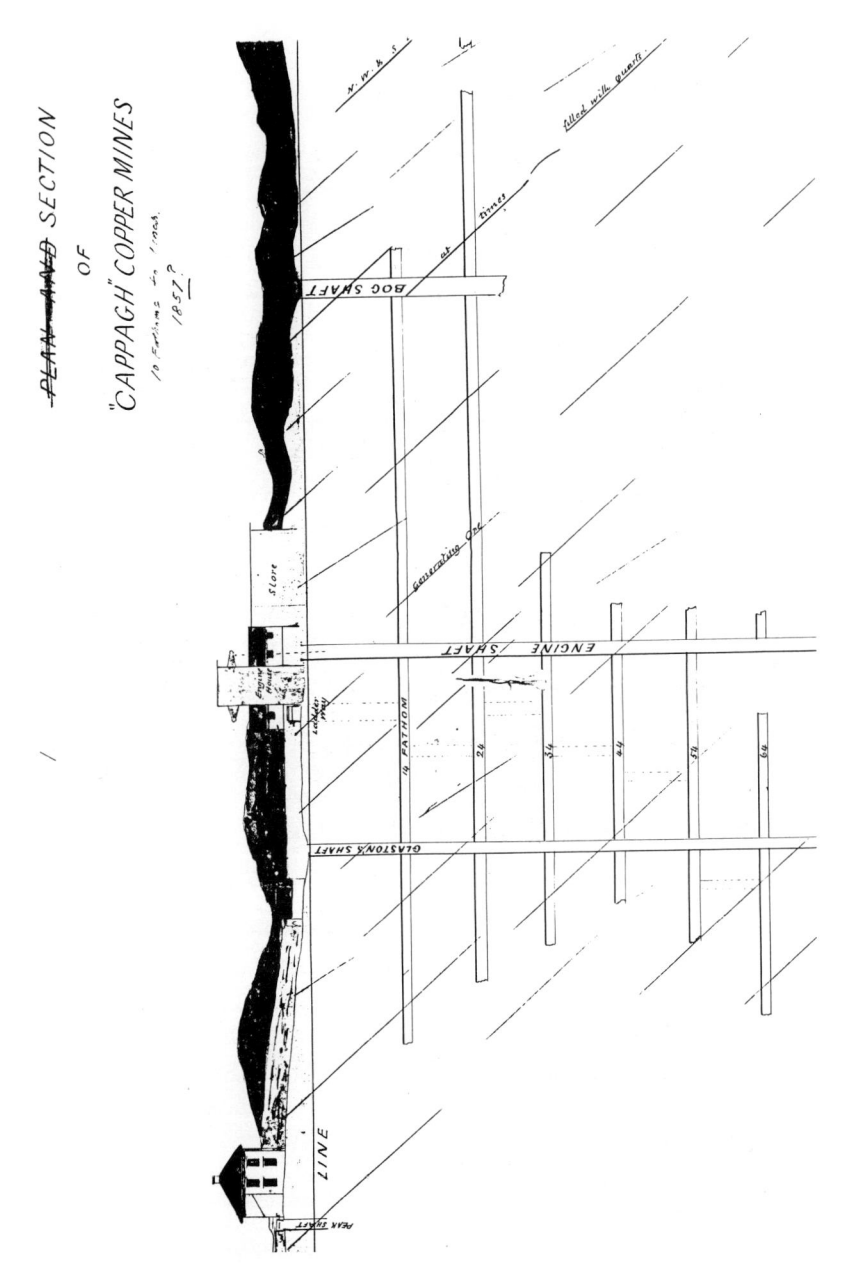

Figure 5. "Cappagh" Copper Mines, Co. Cork

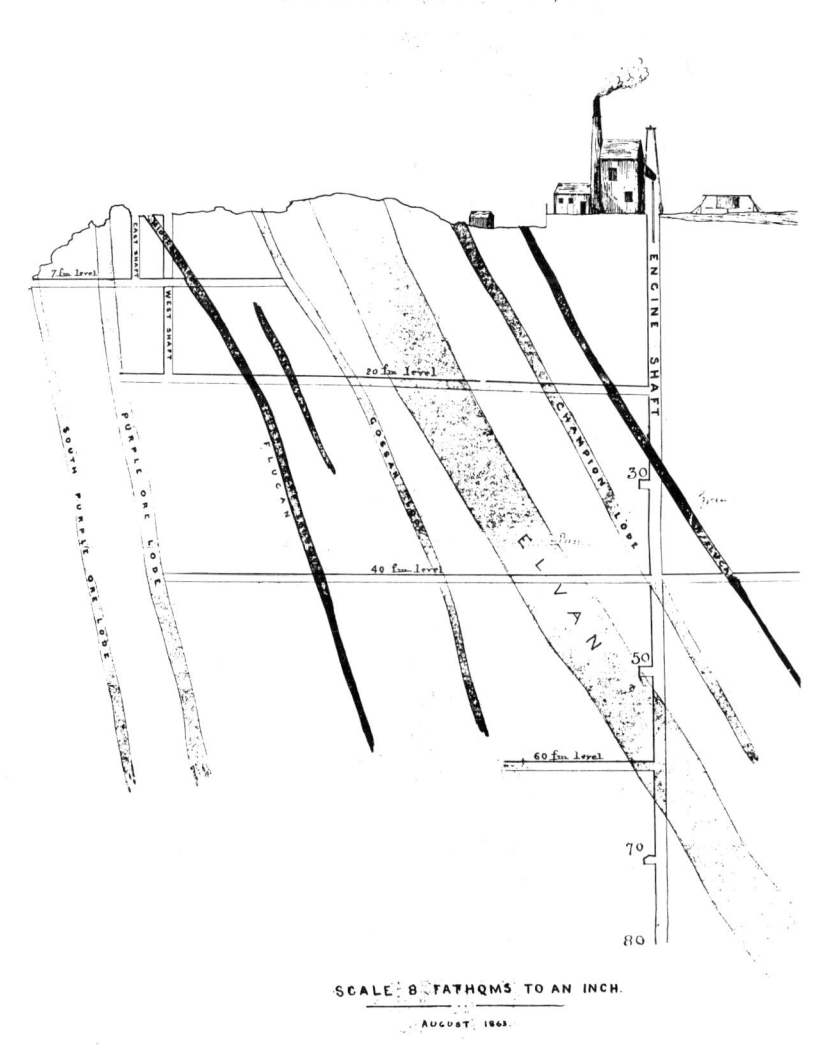

Figure 6. Crookhaven Mines, Co. Cork

Figure 7. Glandore Mines, Co. Cork

Figure 8. Creevelea, Co. Leitrim

Figure 9. Glendalough, Co. Wicklow

Figure 10a & 10b. Conlig, Co. Down

xxx

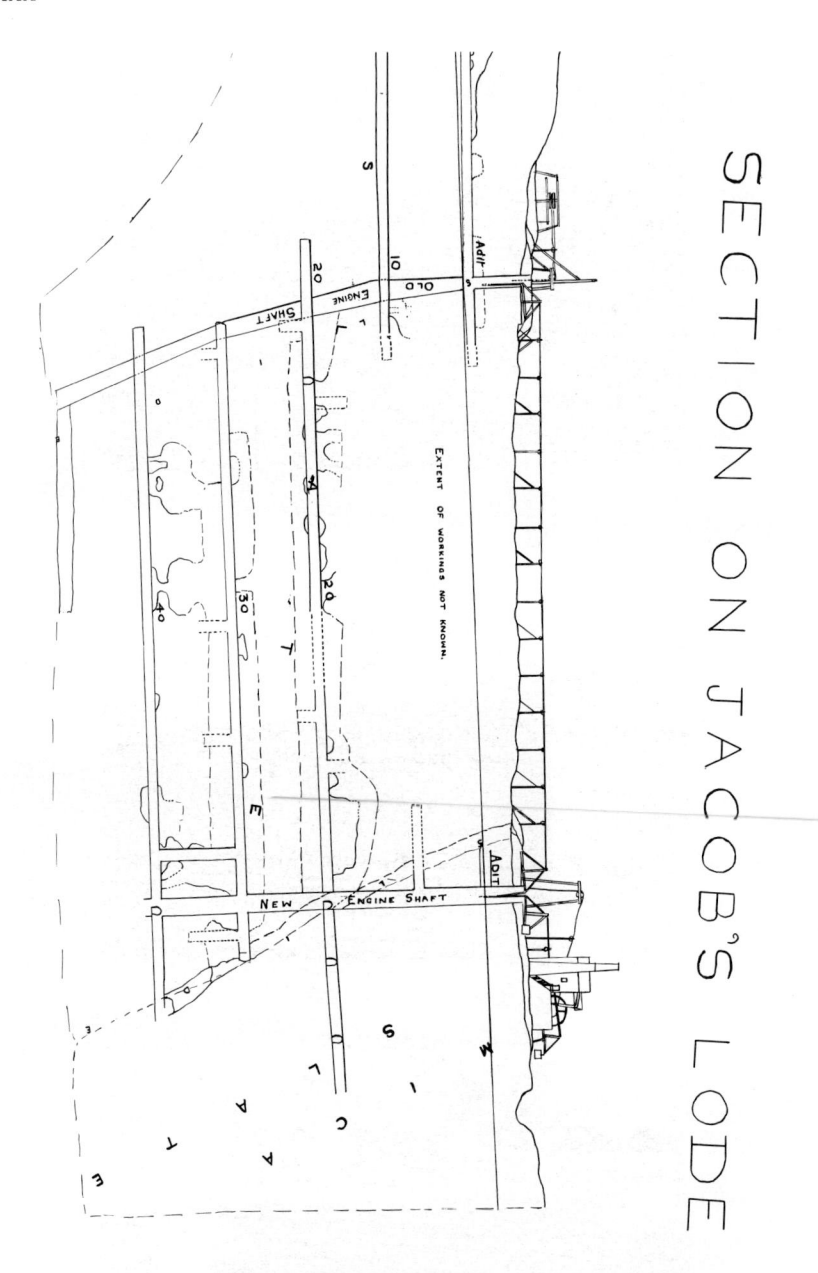

Figure 11. Keeldrum Mine, Co. Donegal

xxxi

Figure 12a, 12b &12c. Conlig/Whitespots, Co. Down

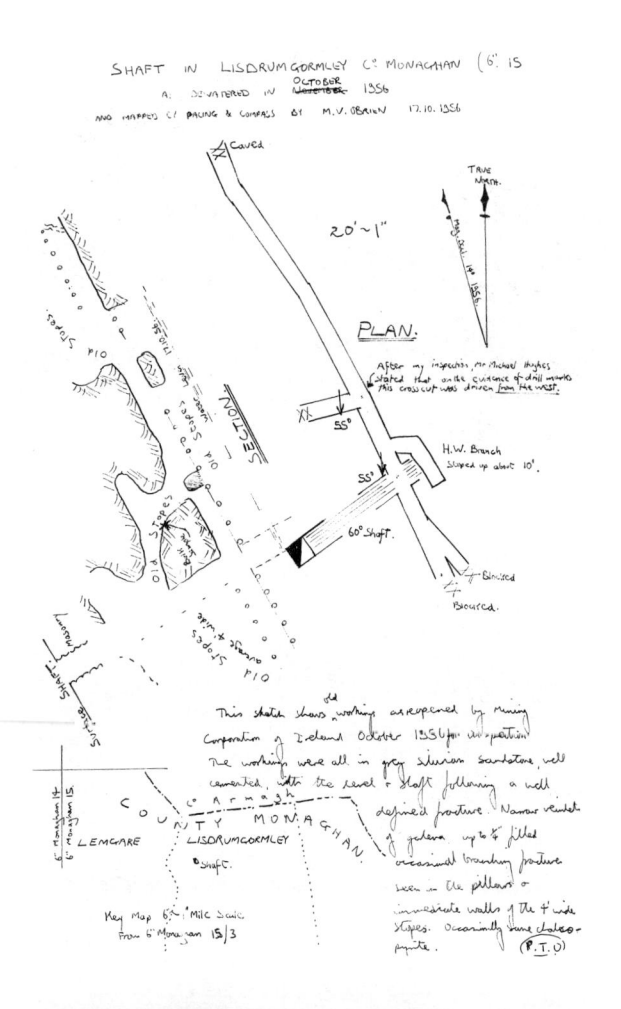

Figure 13a & 13b. Lisdrumgormly, Castleblayney, Co. Monaghan

Figure 14. Glengowla Mine, Oughterard, Co. Galway

Figure 15. Hope Mine, Castleblayney, Co. Monaghan

Figure 16a, 16b &16c. Ballycorus Mine and Leadworks, Co. Dublin

Figure 17a, 17b &17c. Hero Mine, Luganure, Co. Wicklow

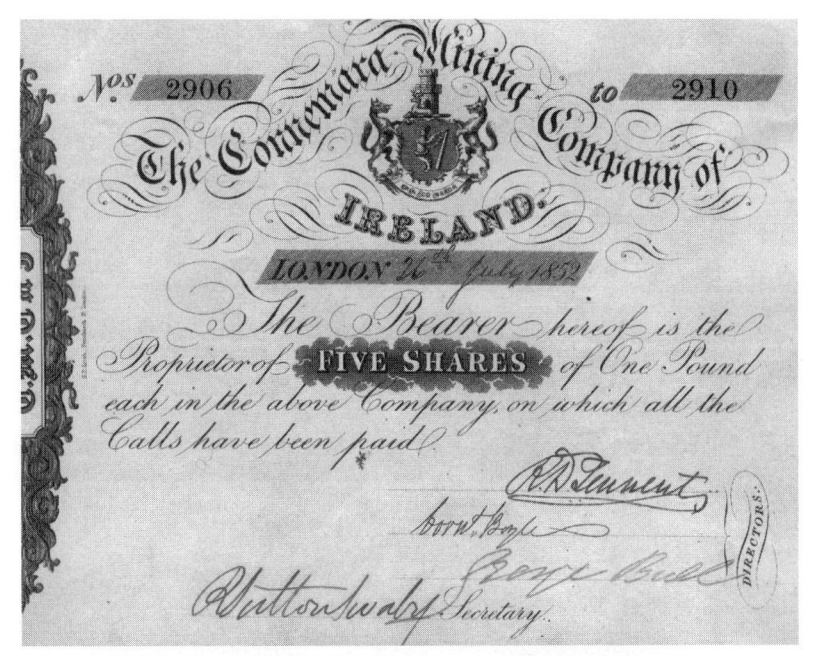

Figure 18. Connemara Mining Company of Ireland share certificate.

xxxvii

Figure 19. Ballyhickey Lead Mine, Quin, Co. Clare

Figure 20. Kilbricken Mine, Quin, Co. Clare

Figure 21. Shallee, Co. Tipperary

Figure 22. Barrystown, Bannow, Co. Wexford

Figure 23a and 23b. Dhurode Mine, Co. Cork

SIR RICHARD J. GRIFFITH, BART.
(1784-1878).

From the bust by Sir T. Farrell, by permission of the Royal Dublin Society.

DEPARTMENT OF AGRICULTURE AND TECHNICAL
INSTRUCTION FOR IRELAND.

MEMOIRS OF THE GEOLOGICAL SURVEY OF IRELAND.

MINERAL RESOURCES.

Memoir and Map of Localities of Minerals of Economic Importance and Metalliferous Mines in Ireland.

BY

GRENVILLE A. J. COLE, F.R.S., M.R.I.A.

DUBLIN:
PUBLISHED BY THE STATIONERY OFFICE.

To be purchased through any Bookseller or directly from
EASON & SON, LTD., 40–41 LOWER SACKVILLE STREET, DUBLIN.

1922.

Price Seven Shillings and Sixpence Net.

ABBREVIATIONS.

The following abbreviations are used in the Memoir as references to the more important publications on Irish mining :—

GRIFFITH, MAP.—" Map of Ireland to accompany the Report of the Railway Commissioners, etc." Edition of 1855.

GRIFFITH, 1855.—" On the Copper Beds of the south coast of the County of Cork," Journ. Geol. Soc. Dublin, vol. 6, p. 195, 1855.

GRIFFITH, 1861.—" Catalogue of the several localities in Ireland where mines or metalliferous indications have hitherto been discovered," Journ. Geol. Soc. Dublin, vol. 9, p. 140, 1861.

G.S.D. Journal of the Geological Society of Dublin (vols. 1 to 10, 1838 to 1864).

HUNT, 1848.—R. Hunt, Mem. Geol. Survey of Great Britain, vol. 2, part 2. Lead ores raised in Ireland from 1845 to 1847, pp. 706 and 709 ; sales of Irish copper ore at Swansea from 1804 to 1847, p. 713.

KANE.—R. Kane, "Industrial Resources of Ireland." 1844 indicates the first and 1845 the second edition.

KINAHAN or KIN.—G. H. Kinahan, " Economic Geology of Ireland," continuously paged edition in one volume, with index, 1889.

M.C.I.—Mining Company of Ireland, and Report for the year named. The figures i and ii indicate the half year covered by the Report.

MEM.—Memoir of the Geological Survey of Ireland explanatory of the sheet of the 1" map bearing the number given.

MINERAL MAP.—The map accompanying the present Memoir, and its counterpart as a separate sheet.

MIN. STAT.—" Mineral Statistics," published formerly by the Mining Record Office at the Museum of Practical Geology, London, and later by the Mines Department of the Home Office. The year quoted is that with which the statistics deal, the date of publication being a year later. The statistics for 1848–52 are in the records of the Royal School of Mines, vol. 1, part 4 (1853). See also HUNT above.

R.G.S.I.—Journal of the Royal Geological Society of Ireland, in continuation of G.S.D. (vols. 1 to 8, 1867 to 1889). In many libraries the papers may conveniently be found in the Scientific Proceedings of the Royal Dublin Society, in which they were published contemporaneously.

STAT. SURV.—The Dublin Society's Statistical Survey for the county named.

STEWART.—Donald Stewart, Report to the Dublin Society, Trans. Dublin Soc., vol. 1, part 2 (1800), separately paged.

WEAVER.—T. Weaver, " Memoir on the geological relations of the east of Ireland," Trans. Geol. Soc. London, vol. 5, part 1, p. 117 (1819), and " On the geological relations of the south of Ireland," *ibid.*, ser. 2, vol. 5, part 1, p. 1 (1838). The date following Weaver's name indicates the paper referred to.

CONTENTS.

CHAPTER		PAGE
I.	Origin and Method of the Memoir and the Mineral Map	5
II.	Antimony	14
III.	Barytes	15
IV.	Bauxite	22
V.	Copper	26
VI.	Felspar (Feldspar)	61
VII.	Gold	63
VIII.	Gypsum	65
IX.	Iron	66
X.	Lead	89
XI.	Manganese	129
XII.	Molybdenum	131
XIII.	Nickel	132
XIV.	Rock-Salt	133
XV.	Steatite	137
XVI.	Sulphur	139
XVII.	Zinc	143
Index		149

MEMOIR AND MAP OF LOCALITIES OF MINERALS OF ECONOMIC IMPORTANCE AND METALLIFEROUS MINES IN IRELAND.

CHAPTER I.

ORIGIN AND METHOD OF THE MEMOIR AND THE MINERAL MAP.

At the outset, attention must be called to the scope and title of the Map accompanying the present Memoir. On it are marked " the principal localities of minerals of economic importance," such as ores of lead, gypsum, and steatite. The resources of Ireland in the form of sands, clays, or marble, are not indicated, nor are they covered by the title of the Memoir. These materials are rocks. They are naturally included in G. H. Kinahan's " Economic Geology of Ireland " (1888), and have been generally touched on by the present writer in " Ireland, Industrial and Agricultural " (p. 18, Department of Agriculture and Technical Instruction, 1901 ; 2nd ed., 1902), and by Mr. E. St. John Lyburn, A.R.C.Sc.I., in " Papers read at the Industrial Conference, Cork International Exhibition " (p. 65, Dept. of A. and T. I., 1903). Their recent development has been dealt with by Mr. Lyburn in an illustrated paper on " Irish Minerals and Raw Materials " (Journ. Depart. of A. and T. I., vol. 16, p. 121, 1916). The anthracite of Kilnaleck appears on the map as an adjunct to the coalfields. The Diatom-earth that extends from Toome Bridge to Portglenone is also indicated ; but its description falls outside the scope of a Memoir that deals with " localities of minerals of, economic importance and metalliferous mines." The emphasis here lies on the word " minerals."

Here, again, a warning note seems necessary. The naming of a locality as containing a mineral of " economic importance " by no means implies the importance of the occurrence at the spot. The title of the Memoir has been carefully chosen, and the detailed descriptions will, it is hoped, dispel any illusions that might be founded on the number of mineral localities entered on the Map.

The Memoir originated in a series of notes that were brought together under the stress of war-conditions, when public and private enquiries showed more than ever the need for a systematic review of the mineral resources of the country. Some thousands of references have now been correlated by the writer, and it is hoped that his selection of the facts of primary interest, aided by an acquaintance with the topographical and geological features concerned, may be of service when any particular mine, townland, or district comes up for consideration in the future. Where so many figures and so much material have been involved, and where it has been often necessary to criticise the accuracy of predecessors, errors on the part of the author have probably crept in. On the other hand, matter that might have been included has possibly remained outside. The writer and his successors will be grateful at all times for additions and corrections.

Mr. E. St. J. Lyburn, as Economic Geologist to the Department of Agriculture and Technical Instruction, has contributed very useful information in regard to recent workings, and Mr. J. de W. Hinch, of the Geological Survey, has helped greatly in clearing up, from the 6″ MS. maps of the Survey, obscurities in the records of the past.

The Map shows the localities where mines, exclusive of individual coal-mines, have been opened at various times, and also a number of places where minerals of economic importance are found in sufficient quantities to attract attention. At the request of the Department of Agriculture and Technical Instruction, the Geological Survey, in 1918, used the Index Map of the Ordnance Survey, on the scale of one-tenth of an inch to one mile (1 : 633, 600), as a basis on which bogs and coalfields were inserted. Under the care of Dr. J. F. Crowley, A.R.C.Sc.I., the railway-lines on this map were emphasised, and the navigable waterways through lakes, rivers, and canals were added in blue. On the "fuel map" thus prepared for the Department, I have now marked the mineral localities, so that their positions on the 1″ sheets of the Ordnance or Geological Surveys can be readily found. The indications on the MS. 6″ sheets of the Geological Survey, and other sources of information, have been freely utilised; but disused mines have a well known way of disappearing at the surface, and in many cases the identification of their shafts becomes a matter of historical research.

This Mineral Map has been reproduced under the care of the Ordnance Survey, and I cannot help recording my sense of the consideration given to its details and to its revision as the work went on. The copies issued with the Memoir are printed with the topography in grey and the names of mineral localities in red. These copies are so folded that four unfoldings expose the whole surface of the map.

An issue of the map on heavier paper with names in green

ORIGIN AND METHOD OF MEMOIR AND MAP 7

has also been published as an unfolded sheet, which should serve as a convenient index to Irish mines, and to the one-inch maps of the Ordnance and Geological Surveys, if hung on the wall of an office, a library, or a school-room. On this issue the topography and sheet-boundaries are emphasised. Since these unfolded copies were printed off first, opportunity was taken to insert on the folded copies the names of eight additional localities, which are not very important in themselves, but which were already referred to in the Memoir.

In each case a coloured dot on the map indicates the site of the mine or mineral locality, and the chemical symbols of the prevalent materials are placed, where possible, near these dots. The symbols used are as follows :—

Al.	Bauxite (Aluminium Ore).
Au.	Gold.
Ba.	Barytes (Barium sulphate).
Cu.	Copper ore (in almost all cases copper pyrites).
Fe.	Iron ore (Haematite, Hydroxide ores, Siderite, rarely Magnetite).
FeS_2.	Iron pyrites (" Sulphur ore ").
Mn.	Manganese ore (various oxides).
Mo.	Molybdenite (Molybdenum sulphide).
Pb.	Lead ore (Galena ; some Cerussite).
Sb.	Antimonite (Antimony sulphide).
Zn.	Zinc ore (Blende and Calamine ores).

In the summary descriptions, the name of the mine or mineral locality is given in thick type ; modified forms, which are often corrupt, are printed in italics. The position on the map of Ireland is shown by the reference to the sheets of the one-inch map of the Ordnance Survey and also to the six-inch county maps. The references are abbreviated thus :—

1″ 200. 6″ Cork, 142 S.E. Each 6″ map is supposed to be divided into four quarters, lettered N.W., N.E., S.W., and S.E. respectively. The occurrences are arranged under the head of the material present, and, as far as possible, in the order of the one-inch sheets concerned. This order brings them into series running primarily from left to right, and secondarily from the top to the bottom of the Mineral Map ; but a group of mines is in certain places conveniently dealt with as a whole, in spite of its extending over sheets that do not follow precisely in this easy sequence.

An excellent index map of Ireland, by counties, on the scale of $\frac{1}{4}$″ to one mile, and showing the boundaries of the 6″ and 1 : 2500 sheets, is published by the Ordnance Survey as a bound quarto volume, price 5s. The surroundings of the mines that are described in the present Memoir can be realised from this index in considerable detail, and the book forms incidentally a complete county atlas of Ireland on a uniform scale.

8 ORIGIN AND METHOD OF MEMOIR AND MAP

The following notes may assist the reader in using the Memoir and the Mineral Map :—

(i) If the name of a mineral locality is found on the Map, the corresponding description will be found from the index to the Memoir.

(ii) If the name is known and its location is required, the 1-inch sheet-number quoted in the description in the Memoir will indicate its position on the Mineral Map.

(iii) From this indication the position of the mine or lode or mineral locality can easily be traced on the corresponding 1-inch map of the Geological Survey. *For roads and other means of access to the locality the latest 1-inch Ordnance map should always be consulted and used when travelling in the country.*

(iv) If localities of ores of particular metals or minerals of economic importance are required, the classified descriptions, grouped under the head of the particular substance, give the names of the localities and references to the 1-inch sheets, and consequently serve as references to the index-map.

It is obvious that the sites marked upon the various maps include mines that have been long abandoned. In some cases, the mineral deposit became worked out, after amply repaying those who undertook its exploitation, and further trials failed to justify additional expenditure. Modern methods of exploration may reveal at these spots bodies of ore hitherto unknown, and the conditions of the metal-market and of transport from foreign lands must determine future investigations. In other cases, the quality of the ore may have been excellent, but its quantity was so vastly surpassed by the output of some district overseas that competition was rendered impossible in the open market demanded by manufacturing industries. In the case of most of the Irish iron ores, smelting became unprofitable in view of the exhaustion of local wood-fuel and of the great developments in connexion with the British coalfields. The local iron-furnaces in the Weald of England suffered from the same causes. It is the custom to attribute the closing of a mine in Ireland to the incompetence of some particular manager or to trade jealousy across the Channel; but those who undertake development in the future will naturally share a broader outlook, and will be acquainted with the history and fluctuations of the mineral industry as a whole.

The cutting off of external mineral supplies may, of course, lead to a notable, if temporary, revival in cases that otherwise could not be regarded as commercial propositions.

The sources of information as to mining in Ireland are very limited until we reach the literature of the eighteenth century. It is common when a lode is being examined to come across " Old Men's workings," long ago forgotten. These may be of very various dates, but are commonly attributed to the Norsemen ("Danes"), whose enterprise, penetrating inland from their civic settlements on the coast, made a lasting impression on the country. The Normans, however, seem to have found few mines in operation, and we may pass at once to the records of the last three hundred years.

Gerard Boate completed his "Ireland's Naturall History" in 1645, depending largely on information procured by his brother Arnold, who spent eight years in the country, and who travelled frequently from Dublin into the provinces. Gerard himself came to Dublin as State Physician in 1649, but died after a few months in the capital in January, 1650. In spite of an obvious bitterness engendered by the rising of 1641, his record appeals to the modern reader as careful, systematic, and scientific. It was brought out under Arnold Boate's auspices in London in 1652; but nothing seems to have been added since the end of 1645. A good reproduction was issued in Dublin in 1726 by Sir Thomas Molyneux (or Molineux), with additional chapters, and this was republished in 1755. The 1755 edition is that usually available. The earlier chapters are numbered as in the rare small octavo edition, but a few changes in spelling were made. Chapter 16, "Of the Mines in Ireland, and in particular of the Iron-mines"; Chapter 17, "Of the Iron-works"; and Chapter 18, "Of the Mines of Silver and Lead in Ireland," contain much information as to mining in Ireland in the seventeenth century.

At the end of the eighteenth century, the enterprising Dublin Society sent Donald Stewart, who made no claim to being a trained mineralogist, on a tour through most of the counties of Ireland, so that his notes might help those authors who were about to prepare statistical surveys for the Society. His report is printed in Trans. Dublin Soc., vol. 1, part 2 (1800). It is very vague, and the localities are poorly indicated. The writer seems to have had a special eye for what he calls "manganese," and a miner's faith in black shales as giving hopes of coal. Here and there the report is useful, as incidentally mentioning the working of a mine, and it brings to light some occurrences otherwise forgotten.

Sir R. J. Griffith, the greatest of the pioneers in Irish geology, began a series of MS. notes on Irish mines as far back as 1821 (G.S.D., vol. 9, pp. 21 and 140, 1861, and Maxwell Close's review of his work, R.G.S.I., vol. 5, p. 136, 1880). He brought to his work experience gained in the mining districts of England and Wales and Scotland; he was a member of the Geological Society of London from the second year of its existence, and had already acted for nine years as mining engineer to the Dublin Society. The Society published his "Report on the metallic Mines of the Province of Leinster" in 1828, as a small volume that has now become very rare. I am informed by the late Registrar of the Society that 750 copies were printed. Two of these are now in the National Library of Ireland; but few others seem to be forthcoming.

A number of notes on mineral occurrences are included in the Statistical Surveys of twenty-three counties of Ireland, published by the Dublin Society (from 1820 Royal Dublin Society) in the first thirty years of the nineteenth century. T. Weaver, who was engaged on mining operations in the Wicklow district, made memorable contributions to our knowledge of Irish metalliferous lodes in two papers read before the Geological Society of London in 1818 and 1835 ("Memoir on the geological relations of the east of Ireland," Trans. Geol. Soc. London, vol. 5, part 1, p. 117, 1819, and "On the geological relations of the south of Ireland," *ibid.*, ser. 2, vol. 5, part 1, p. 1, 1838). In 1838, the Railway Commissioners published the first edition * of Griffith's "General map of Ireland, showing the principal physical features and geological structure of the country," a work on the scale of ¼ inch to 1 mile, replacing his smaller maps already issued. About this time the Ordnance Survey was publishing engraved 6″ and 1″ sheets, on which a number of mines and their associated works were indicated. It is always worth while to examine the early editions of these sheets, when current editions have ceased to furnish information.

Sir R. Kane recorded a number of original observations in his well known "Industrial Resources of Ireland" (ed. i., 1844; ed. ii., 1845), which will always remain a classic work of reference. Its propositions, however, are often quoted by enthusiasts as if the economic conditions of the middle of the nineteenth century still prevail in the mining industry of Ireland.

In 1853, Sir W. W. Smyth's authoritative memoir "On the Mines of Wicklow and Wexford" appeared as part 3 the first volume of the Records of the Royal School of Mines; this can still be purchased in a separate form. Smyth, it may be mentioned, pays a fine tribute on p. 405 to the personality of T. Weaver, whom he knew.

Joseph Holdsworth published in 1857 a small work entitled "Geology, Minerals, Mines, and Soils of Ireland," which is quite unsystematic, but which brings a few of Kane's descriptions down to a later date. Many mines are mentioned by Holdsworth under unusual names, so that identification is difficult; the spelling of place-names is, moreover, unsatisfactory. Since the work has no index, it may be well to mention that the most interesting references to mines are on pp. 23, 24, 29, 38, 39, 41, 47, 59, 69, 70, 72, 85–88, 102, and 119.

By this time, the Geological Survey, under its local director, J. Beete Jukes, was producing with remarkable celerity its

* This map had been exhibited in successive MS. forms before this date and these have been informally called editions by M. H. Close and others.

1″ maps and accompanying memoirs. The first dated sheets are those of the country near Bray, Baltinglass, and Wicklow town, issued in 1855. The original MS. lines and field-notes on which these publications are based are set down on the Ordnance Survey 6″ sheets preserved for public reference in the Geological Survey Office in Dublin. Very many mineral lodes were engraved and coloured by gold lines on the published 1″ maps; and the ore present in each case was indicated by its conventional symbol. The names of mines, however, were not systematically inserted, and those that appear on the sheets seem to be due to the earlier work of the Ordnance Survey. The Memoirs of the Geological Survey of Ireland have commonly included notes on mines and minerals, and in several cases plans and sections of mines have been inserted. The subject, however, was held to lie rather in the domain of the Mining Record Office, and the attention given to it naturally varied with the personal bent of the geologist concerned in the preparation of the memoir. It is noteworthy that questions of metalliferous ores and their older mining history were avoided, evidently by set purpose, in J. E. Portlock's important memoir for the Ordnance Survey in 1843 ("Report on the Geology of Londonderry," etc., Dublin and London). A card catalogue is now being compiled, county by county, from the notes written on the 6″ Geological Survey sheets, so that records can be referred to without searching over the actual MS. maps. As is stated in the printed list of the publications of the Survey, manuscript copies of the original 6″ geological maps can be supplied at a charge covering the cost of drawing and colouring and of the 6″ sheet.

In 1855 the definitive edition* of Griffith's great map was published, and uncoloured copies of this, with all its geological indications, are still procurable from the Ordnance Survey. This completely revised edition contains the names of a large number of mines and mineral occurrences not to be found on any other published map. The delicate engraving of the names and mineral symbols makes it possible to overlook on the coloured copies some of this exceptionally valuable information. Shortly afterwards Griffith furnished what is virtually an index to the map in his masterly "Catalogue of mines or metalliferous indications," which forms part of his paper on "The localities of the Irish Carboniferous fossils . . ., with the Irish mining localities . . . originally compiled for the use of the General Valuation of Ireland," G.S.D., vol. 9, p. 140, 1862 (the part was published in 1861). This paper may also be found in the "Dublin Quarterly Journal of Science" for 1861, p. 244. The names are conveniently grouped under

* Mr. T. Sheppard ("Nature," vol. 106, p. 243, 1920, and letter to the present writer), who is well known to geologists for his bibliographical studies, has found in the library of the Geological Society of London a copy of an issue with the engraved statement, "The Geology revised and improved in 1853." This edition seems to have been soon superseded.

counties, and references are given to the 6″ sheets in which the localities may be found.

G. H. Kinahan, of the Geological Survey, repeated and somewhat modified Griffith's list of mines in his "Geology of Ireland," p. 339 (1878), and from 1886 to 1889 he published, through the R.G.S.I., under the title of "Economic Geology of Ireland," a series of records arranged under mineral headings and also under counties. From his extensive observation and experience, he was able to add greatly to the material that he gathered from the Memoirs of the Geological Survey, which contain many of his own contributions, and from other sources, including oral tradition. Kinahan's work remains available as a single volume of 514 pages, and will also be found in libraries in the form in which it was originally published, as parts 1, 2, and 3 of volume 8 of the Journal of the Royal Geological Society of Ireland (1889), or, identically printed, in the Scientific Proceedings of the Royal Dublin Society, volumes 5 and 6 (1886–91). In the spelling of names and technical terms, Kinahan's work is often in need of press-correction.

Robert Hunt, in his "British Mining," 1884, gives some account of Irish mines, as on pp. 275–282, pp. 468–475, etc.; but his sources of information do not seem to have gone beyond what had already been published by Weaver, Kane, Smyth, and other authors whose work is utilised in the present Memoir.

G. H. Kinahan's paper, "Notes on Mining in Ireland" (Trans. Inst. Mining Engineers, Newcastle-on-Tyne meeting, 1904), adds nothing fresh to what this author had already published.

The "Report of the Controller of the Department for the development of Mineral Resources in the United Kingdom" (Stationery Office, Cd. 9184, 1918) contains references to visits in recent years on behalf of the Department to Irish mining localities (pp. 21, 41, 54, etc.), and a summary of present conditions in regard to Irish mining (p. 48). The Reports furnished by officers of the above-mentioned Department on certain Irish mines are lodged in the office of the Imperial Mineral Resources Bureau, Mines Department, London.

With regard to the statistics of output of Irish mines, the sales of copper ore at Swansea for the years 1806–1847 are given in a table in the "Memoirs of the Geological Survey of Great Britain," vol. 2, part 2, p. 713 (1848). The output of lead from Irish mines is given for 1845–7 in the same volume, pp. 706 and 709 These returns are continued for 1848–52 in the "Records of the School of Mines," vol. 1, part 4 (1853). The Geological Survey of Great Britain published "Mineral Statistics of the United Kingdom" from 1855 to 1880, including the years 1853 to 1879, and these annual octavo volumes were continued as Parliamentary Papers in a folio form by the Home Office from 1880 onwards. The local names used by mining companies frequently differ from those given on Ordnance Survey maps; the latter have been taken as the only practical standard in the present Memoir. In the official tables and the Mineral Statistics of earlier years confusing misspellings occur, no attempt having been made to check the names or to maintain uniformity from year to year. The same mine, moreover, is placed under various counties, and in some cases it has been impossible to trace the locality referred to. The indifference of even the owners of mines to systematic nomenclature and spelling is, no doubt, responsible for some of this confusion. The summer-schools organised by certain universities for Library Assistants might well form a missionary branch for mining engineers.

Where a mine has no present output, its name may none the less be found in many cases in the " List of Metalliferous Mines in the United Kingdom," published in the Mineral Statistics from 1859 to 1880, and since as the annual " List of Mines in the United Kingdom of Great Britain and Ireland " (Home Department). The first list in this latter series was issued in 1884. The names of proprietors and managers recorded in these lists form a useful guide to the history of a mine.

The Annual Reports of the Mining Company of Ireland, which are now very difficult to procure, have been kindly lent to us by Mr. Joseph Dobbs, of Milltown, Co. Dublin; these have supplied many valuable details of workings in the nineteenth century.

The present Memoir is, on the face of it, a compilation of dry and disconnected facts; but it has been written with a continuous outlook across the country as a whole. The grouping of the mineral occurrences in certain districts raises considerations of geological structure that carry us back to the remote history of the earth. The development of these occurrences as mines gives us an encouraging picture of human enterprise, whether among the well regulated lowlands, traversed by roadways and canals, or in difficult uplands and the recesses of stream-dissected moors. In many cases the treasures of the rocks have been successfully carried off by pioneers; in others much may still remain to justify future methods of exploitation. The details here brought together may at any rate direct attention from the less profitable fields to those where industrial successes may yet be won. Though mining industry may have first been stimulated by the demand for swords of bronze and spears of iron, the need for metals in the arts of peace is happily continuous and increasing.

CHAPTER II.

ANTIMONY.

In a few Irish lead mines the ore is recorded as antimonial; but only in two cases has antimony been the actual cause of mining. The mineral sought for is antimonite (stibnite), Sb_2S_3, containing 71·8 per cent. of antimony.

Castleshane. 1″ 58. 6″ Monaghan 10 S.W. A trial was made here, 3½ miles east of Monaghan town, by the M.C.I. in pursuit of antimony ore (Rep. 1826, ii.). A shaft was put down, but the Company did not think the indications satisfactory (Rep. 1826, iii.).

Clontibret Mines. 1″ 58. 6″ Monaghan 14 N.E. The antimony ores referred to under this name were mined in a shaft in **Tullybuck** and two shafts in **Lisglassan** (*Lisglassin*). Mem. 58, ed. 2, p. 21 (1914), gives a map showing their relative positions. The church on this map is ½ mile N.N.W. of Milltown, on the road to Monaghan town. The antimonite is associated with galena (see also under Lead). Stewart (p. 101) claims to have discovered an antimony "course" near "*Glentubert*" church—that is, the church south of Milltown—in 1774. Since he says that it was in a stream-hollow, it is probably the vein worked later in Tullybuck, where Griffith (1861, p. 150) says that Lord Midleton opened a mine. Stewart (p. 102) speaks of a lead and antimony mine in "*Lisglassin*" as distinct, and existing in 1800; but the word "mine" in old usage often refers to a mere unworked body of ore. Kinahan (p. 150) writes of "Clontibret" as if if were the name of a third separate mine; but of this there seems no record.

Kane (1845, p. 233) says that the Clontibret vein (which he places in Armagh) was "about four inches thick, with a bedding of quartz" in the Silurian slates. One shaft was 7 fathoms deep, with several levels; but the enterprise reaped little profit. The mines were reopened under Government advice in 1917 (Rep. of the Controller of the Department of Min. Resources, 1918, p. 40), and in the Home Office List of Mines for 1918 Mr. Robert Espinasse, of Dundalk, is named as the owner.

Connary. 1″ 130. 6″ Wicklow 35 N.W. As will be also noted under Lead, Weaver (1819, p. 215) describes the mixed lead and zinc ore of Connary as including antimonite. This mineral, however, does not seem an important constituent of the "bluestone," which was raised by the Ovoca Mineral Company down to 1885. P. H. Argall ("Mining operations in the East Ovoca district," R.G.S.I., vol. 5, p. 160) uses the name "kilmacooite" for this mixed material, and says that it contained antimonite in Kilmacoo, which is a working on the Connary lode just N.E. of Connary. The total quantity of antimony must be small, since it does not figure in the four analyses of "kilmacooite" given by Argall on p. 164, nor was it found by Apjohn in 1851 (G.S.D., vol. 5, p. 134, 1852). For the "bluestone" see also Connary under Zinc and Lead.

Kilbreckan (*Kilbricken*). 1" 133. 6" Clare 34 S.W. This mine will be more fully described under Lead. It was opened in 1834. Apjohn (Proc. R.I. Acad., vol. 1, p. 469, 1840) detected a mineral from the lode with the following analysis: S 7·25, Sb 6·37, Pb 30·52, Fe 0·17, Loss 0·2 (Total 44·52), and named this joint sulphide kilbrickenite. J. D. Dana ("System of Mineralogy," p. 105, 1868) shows it to correspond with geocronite, which was named by Svanberg in 1839. He calculated Apjohn's analysis to a total of 100 as follows:— S 16·36, Sb 14·39, Pb 68·87, Fe 0·38.

Ballinvirick (Ballintredida). 1" 143. 6" Limerick 20 N.E. This lode has become lost to sight (see under Lead later); but Weaver (1838, p. 65) records among its contents antimonial galena and oxide of antimony, and also "antimonial copper ore," presumably tetrahedrite.

CHAPTER III.

BARYTES.

Since the Geological Survey of Ireland proposes to issue a special memoir by Mr. T. Hallissy on the deposits of barytes (barium sulphate), little more than a list of the more important occurrences need be given here, with a few notes on their development and history. The association of this heavy white mineral with a large number of veins of lead and copper ore was well recognised in the eighteenth century; but commercial uses for barytes, some of them at first of a dubious nature, are a discovery of comparatively recent years. Weaver, for instance (1838, p. 61), mentions the veins at Shallee near Silvermines, but only as a guide to lead ore.

Baryta is the name of barium oxide, corresponding to magnesia, alumina, and the other "earths" of older writers. The name Barytes was also used for the "earth" (W. Phillips, for instance, "Outlines of Mineralogy and Geology," p. 14, 1816), and hence the mineral was called "sulphate of baryta," and "sulphate of barytes," indifferently; Weaver employs the latter form. The mineral itself became called Barytes by English writers about 1830, and J. D. Dana, at a later date, introduced Barite, following the use of Baryte by Haüy in 1801. Barite is thus the common name of the mineral in American literature.

The barytes that fills veins and fissures, and sometimes forms great masses, descending like pipes into the depths, seems to have been deposited by chemical interaction where it is now found. A. M. Finlayson ("Ore deposition in the lead and zinc veins of Great Britain," Quart. Journ. Geol. Soc. London, vol. 66, p. 308, 1910) points out that the barium may have been brought up in solution in water as a bicarbonate; Lattermann and Hardman (G.S.I., vol. 5, p. 103, 1878) have suggested that it was in the form of chloride. In any case, the associated sulphates, by their oxidation, have caused its deposition as the almost insoluble sulphate, the mineral barytes.

The development of the barytes trade in Ireland will be referred to under the head of the Bantry district.

Gleniff. 1" 43. 6" Sligo 6 S.W. Gleniff has been cut by a stream running to the Duff River, on the north side of the high plateau-mass of Ben Bulben. Across the glen-head, and passing through the mountain south-eastward to Glencar, a great lode of barytes has been traced in the Upper Carboniferous Limestone, forming a conspicuous feature on the 1" geological map. It is in one place 5 feet and in another 6 feet wide, and its connexion in depth with mineral sulphides is evidenced by the occurrence of traces of copper pyrites and galena.

"Gleniff Barytes" appears in the List of Mines in Min. Stat. 1875. It was worked by Sir R. Gore Booth, and by Mr. Barton in 1878-79 (Mem. 42 and 43, p. 29, 1885). Kinahan (p. 39) mentions the mine as "Glencarberry" (**Glencarbury**). The first record of output is in Min. Stat. 1890, where the Gleniff mine appears as a contributor to the total of barytes raised in Ireland. A separate output of 638 tons is stated for 1894; 300 tons are recorded for 1896, and 1,350 for 1897. Since then the average has been 450 tons annually (378 tons in 1916, given under Bundoran, in the Report of the Department of Min. Resources, p. 27, 1916). The mine was worked in 1918 by the Gleniff Barytes Company, Glasgow.

Tormore (Glencar). 1" 43. 6" Sligo 9 N.W. The south-eastern end of the great vein of Gleniff penetrates the cliff-wall of Glencar, and is worked by an open cut in Tormore, high up on the mountain side and close against the Leitrim border. In 1915, the barytes was being extracted by the Sligo Barytes Company, of St. Stephen's House, Westminster, London.

Carrickartagh. 1" 70. 6" Monaghan 27 S.E. Barytes was raised in this townland, near Carrickmacross, about 1898, and the deposit is again being worked by the Farney Development Company, who re-opened the shaft in 1919. The lode occurs in the Silurian strata of the district.

Cloosh (Clooshgereen). 1″ 105. 6″ Galway 54 S.W. An account of this mine is given under Lead. A plan and section are in Mem. 105 and 114, p. 59 (1869). The gangue is mostly barytes. A similar lode, also worked for lead, is in the adjacent townland of **Cregg.** The output of 149 tons of barytes quoted from Oughterard in Min. Stat. 1882 as raised by Mr. H. E. A. Young of Galway, was probably from one or other of these mines. No other record from the Oughterard district exists.

Baravore (*Barravore*). 1″ 130. 6″ Wicklow 23 S.W. This is one of the Glenmalur mines (see under Lead). The lode contains an important quantity of barytes.

Silvermines Mines. 1″ 134. Barytes is conspicuously associated with the lodes in this district. See Mem. 134, pp. 19, 37 and 45, and Weaver 1838, p. 61.

Killiane. 1″ 169. 6″ Wexford 43 S.W. This townland is 4 miles south of Wexford town, and barytes lodes occur in the area between it and the South Intake of the harbour, the largest being 5½ feet wide. Specks and strings of galena are noted on the MS. 6″ map of the Geological Survey. See also Mem. 169, etc., p. 52 (1879). Kane, who did not note the barytes of Duneen and West Carbery, mentions the Killiane lodes (1845, p. 245). A trial was put down on one of these before 1879.

The Bantry District.

Barytes is very common as the gangue of lodes of lead and copper sulphides throughout the mining region of Co. Cork. There is, unfortunately, much difficulty in interpreting the official records of material exported from the neighbourhood of Bantry. Of the known local mines, Derryginagh, east of the town, is probably responsible for the early outputs registered as "Bantry" and "Bantry Bay," though something may have come from the mine near Mount Gabriel, north of Skull, which Triphook (G.S.D., vol. 6, p. 220, 1856) found at work in 1854. (See Derreennalomane later.) The earliest recorded output of barytes from Ireland is in Min. Stat. 1855, when 1,291 tons were raised by the Barytes Company of Ireland, a quantity double that produced by all other localities in our islands put together. In 1857, "Bantry" produced 700 tons; but by that time Derbyshire was awake to the demand and raised 9,000 tons. In 1862, the next Irish record, Ireland is quoted as yielding, from unspecified mines, 450 tons, and Derbyshire 6,425 tons. Irish barytes then disappears from the records for ten years. Min. Stat. 1874 gives the figures for 1872-4, culminating in 1,506 tons for

1874, valued at £750. A great rise took place in 1875 (15,549 tons, valued at £14,089 10s.). Part of this is probably due to the development of Duneen Bay, since this mine is separately mentioned in the following year as yielding 5,400 tons. "Bantry Bay" remains in the records until 1878, and about this date it becomes possible, despite a variety of spellings, to follow to some extent the fortunes of the separate mines. We must remember that in these records of output "Derryganagh" stands for Derryginagh. Scart is used correctly as a synonym for Derreengreanagh, and the variant "Scart Bay" is probably an invention to harmonise with Duneen Bay. The real trouble is with the name "Derrygranagh," which passes finally into "Derrygrenach." Correspondence with former workers of the mines lead me to conclude that both these names represent, not Derreengreanagh, as might reasonably be inferred, but Derryginagh.

Joseph Dickinson, in his Report as Inspector of Mines in 1881, names a mine near Bantry as "Kilmocapogue"; it was worked by the Bantry Bay Barytes Company, and abandoned in 1881. Mr. Hallissy has ingeniously traced this name to the parish of Kilmocamoge, at the head of Bantry Bay. "Kilmacommoge" appears as a village, the modern Kealkill, on the ¼" map of the Railway Commissioners, which was utilised by Griffith as a basis. The outputs quoted from "Kilmocapogue" are, 1882, 755 tons; 1883, 280 tons. These do not agree with the statement of its abandonment in 1881. This mine was possibly identical with Derryginagh.

I have failed to identify the "Durrus" mine of Min. Stat. 1882, 1883 and 1886. Derreennalomane was worked about that time by the Durrus Barytes Company of London; but it is quoted under its own name in 1884 and 1885. "Durrus" raised 150 tons in 1883, its only recorded output. In that year it is styled "Durrus, formerly Bandon," the latter name being probably derived from Lord Bandon, who worked slate-quarries at Rossmore on Dunmanus Bay in 1864.

Derryginagh (*Derryganagh, Derriganagh, Derrygranagh, Derrygreenagh, Derrygrenach,* probably also *Kilmocapogue*). 1" 192 (extreme S.E.). 6" Cork 118 N.E. A lode, in places 15 feet wide, occurs in Old Red Sandstone in the townland of Derryginagh Middle, 2½ miles east of Bantry. (Mem. 192 and pt. of 199, p. 46, 1864.) Specular haematite occurs on its margins. It was worked shortly before 1864, and then for a time abandoned. E. T. Hardman ("On the barytes mines near Bantry," R.G.S.I., vol. 5, p. 99, 1878), visited Derryginagh, when it was being worked by the Bantry Bay Barytes Mining Company (Min. Stat. List of Mines 1877-9), and he records an output, "when busy," of 20 tons a day. Min. Stat. give only "Bantry Bay" about this time in their records of output, and "Kilmocapogue" in 1882 and 1883.

As above stated, these probably represent Derryginagh. In 1885, "Derryganagh" is credited with 1,500 tons. The output is then joined with those of other mines down to 1893. In 1894, Derryginagh produced 1,125 tons, but only 630 in 1896. The record ceases until the name "Derrygranagh" appears in 1907, probably representing Derryginagh, since Derreengreanagh mine is stated to have been closed in 1906-8, and to have then been yielding very little (letter from Mr. A. F. Storer, of Newcastle-on-Tyne, to Geological Survey of Ireland, 1920). "Derrygranagh" is grouped with other mines down to 1913, and is no doubt responsible for part of the Cork output in 1914 and 1915; in the latter year it is called "Derrygreenagh (Bantry)." The Report of the Controller of the Department of Mineral Resources for 1916, p. 27, gives an output of 2,927 tons from "Derrygrenach," and this name appears in the Home Office List of Mines for 1918, as that of the mine now worked by the Cookson Barytes Company of Newcastle-on-Tyne. These variants of the name of the Derreengreanagh (Scart) lode S.S.E. of Bantry seem thus to have become permanently attached to Derryginagh. Both lodes were formerly worked by the same firm, the Liverpool Barytes Company. The Derryginagh mine is marked as "Barytes Mines" on the 1″ topographical sheet 192 issued by the Ordnance Survey in 1903.

Scart (**Derreengreanagh**; not *Derrygranagh* or *Derrygrenach* of various reports). 1″ 199. 6″ Cork 118 S.E. The Geological Survey about 1863 mapped a lode, which was already known to Griffith, in the northern part of Derreengreanagh and Ardrah townlands, 1¼ mile S.S.E. of Bantry. It figured then as a copper lode, and the copper symbol appears on the Geological Survey 1″ sheet issued in 1859. The reference to Scart in Min. Stat. Lists of Mines, 1862-5, as a lead mine seems erroneous.

The occurrence is described on the MS. 6″ map as a large vein 4 to 6 feet wide of barytes with copper carbonate and specks of copper pyrites. It lies near the crest of the Old Red Sandstone anticline of Knocknareagh, ¼ mile S.E. of hill 636 on the 1″ topographical map issued in 1906.

The barytes proved to be the valuable material in the lode, and the Scart mine was opened here about 1876 by the Scart Barytes Mining Company. On Hardman's visit (R.G.S.I., vol. 5, p. 100, 1878) the pipe of barytes, 30 feet long and 15 feet wide, had already been excavated to a depth of 90 feet. The copper ores formed a very trifling impurity. Two hundred and one tons of barytes were raised in 1876, 1,150 in 1879, but only 91 in 1881, valued at £66. Scart was again working from 1890-3. Mr. A. F. Storer (see above under Derryginagh) informs us that the Liverpool Barytes Company prospected the lode, but closed work there in 1906-8.

The Home Office series of plans of abandoned mines contains one of Scart, deposited as "exhausted" in 1910 (No. 5420). As shown under Derryginagh, records from Derrygrenach and similar names refer to Derryginagh.

Durrus. One of the mines near Bantry was known to Min. Stat. by this name in 1882, 1883 and 1886. It was not Derryginagh, since "Kilmacapogue," which seems to represent that mine, is separately named in 1882 and 1883. The only record of output is 150 tons in 1883.

Derreennalomane. (The second n is often omitted, as has happened in the map accompanying the present Memoir. Griffith, as usual, gives the name correctly.) 1″ 199. 6″ Cork 130 S.E. A considerable lode, with the easterly trend so common in this district, occurs in the Upper Old Red Sandstone on the southern slope of Mount Corin, about 1 mile N.W. of the village of Derreennalomane (Portiroe of the 1″ geological map). A copper lode occurs a little to the south; both are marked on the geological map, and Griffith, on his Map, indicated Derreennalomane as a copper locality. A second copper lode, containing bornite, is cut by the main lode of barytes. If, as seems almost certain, this barytes lode is that described by T. Dawson Triphook ("On the occurrence of sulphate of barytes in the S.W. of the County of Cork," G.S.D., vol. 6, p. 218, 1856), it gave rise to what is probably the earliest barytes mine in Ireland, and, indeed, one of the earliest in the world. Triphook speaks of visiting underground works and two shafts in 1854. In Min. Stat. Lists of Mines, 1862 to 1878, the mine appears as "Brandon Barytes" and *Dreenalamane*, and as worked by Martyn Dennis and Company for copper and barytes.

The official record of output for Derreennalomane begins in Min. Stat. 1854, p. 81, with 2,500 tons, recorded from "Derrenala," the other three localities, all in Britain, yielding a total of only 803 tons. The output soon fell off to 700 tons. One thousand tons are recorded for 1884 and 800 for 1885.

In 1887 the Durrus Barytes Company of London are given as the workers. There is no record from 1890-96, but the mine is named from 1897-98 and in 1899. It appears to have been steadily at work from 1903 to the present time, the output being grouped with that of the other mines in Co. Cork. The Report of the Department for the development of Mineral Resources for 1916, p. 27, records 6,339 tons from Derreennalomane. It was latterly worked by the Dunmanus Bay Barytes Company, 24 Finsbury Square, London, and the product was styled "Dunmanus white." This Company was wound up in 1920. A view of the mine is given by E. St. J. Lyburn in his paper in Journ. Depart. Agric. and Tech. Instr., vol. 16, 1916 (fig. 2).

Dunbeacon. 1″ 199. 6″ Cork 130 S.E. There seems to

have been a mine in this large townland west of the Derreennalomane lode, worked by the British Barytes Company. (Lists of Mines, 1872-4.)

Mount Gabriel. 1″ 199. 6″ Cork 139 N.E. A lode, striking about E.S.E., runs across Mount Gabriel townland, between the wood and the quickly rising crest of the hill, where "Barytes Mines" are marked on the 1″ Ordnance Survey map of 1906. It was not noted on the Geological Survey maps, though Kinahan, in Mem. 200 etc., p. 23 (1861), indicates the occurrence in the road-section to the east of the summit, associated with traces of copper-ores. The mine is nearly ¼ mile W.N.W. of the summit of Mount Gabriel, and three lodes have now been traced. Its record begins with 400 tons in 1894; 1,500 tons were raised in 1896, 200 in 1903, and 295 in 1904. In other years the output is grouped with that of Derreennalomane and Duneen Bay. Mount Gabriel is not separately named after 1907. See also under Copper.

Skeagh. 1″ 199. 6″ Cork 139 S.E. This is a new mine (1918), in Skeagh townland, 3 miles E.S.E. of Mount Gabriel and due W. of Skull. It was opened by Messrs. Smythe, Davis and Tierney, of Dublin. (List of Mines, 1918; for "Scarthe" read "Smythe.")

Letter. 1″ 199. 6″ Cork 140 N.W. This townland adjoins Mount Gabriel on the east. A lode here, marked on the 6″ MS. map of the Geological Survey, was worked from 1889 to 1893, but no separate outputs are quoted.

Duneen Bay. (Duneen; Mukruss Head; *Muckross Head*). 1″ 201. 6″ Cork 135 S.E. and 144 N.E. These mines, south of Clonakilty, have played an important part in the Irish barytes industry, fully comparable to that of the Bantry series. The village of Duneen stands on Duneen Bay, a small inlet between Muckruss Head on the north and Duneen Point on the south. The main barytes lode is shown on the MS. 6″ map of the Geological Survey as ⅓ mile N. of Duneen. Weaver (1838, p. 25) noted three "beds" of barytes in the slaty rocks on the strand of "*Doneen*," the northern being 1½ to 2 feet wide, the middle one about 4 feet, and the southern 6 or 8 ins. Only a few specks of iron and copper pyrites and of galena occurred in them. On the south side of the inlet, as will be mentioned under Copper and Lead, old mining works were visible, which had followed the ores to 30 or 40 fathoms. Griffith, in his Map, noted lead and copper at Duneen; but the locality has since become famous for barytes. In 1876, 5,400 tons were raised, and the total output from 1876 to 1887 inclusive was 60,922 tons, an average of over 5,000 tons a year. From 1888 to 1893 there are no published separate returns; 12,506 tons were raised in the three years 1894-6. After more than 3,000 tons in 1901, the produce was only 72 tons in 1902, and the record is not continuous; but in 1910 and onwards the mine is again included in the official returns with others in Co. Cork. The Report of the Department of Mineral Resources for 1916 gives an output for that year of 4,871 tons. **The Mountain Mine,** in the townland of Mountain Common, a little inland from Duneen, appears in the List of Mines for 1915, and its output is given as 814 tons in the Report above cited. Duneen is now worked by the Cookson Barytes Company of Newcastle-on-Tyne. A view of the Duneen Bay mine is given by E. St. J. Lyburn in his paper in Journ. Depart. Agric. and T.I., vol. 16, 1916 (fig. 1).

Antrim Mines. These appear in Min. Stat. 1878 with an estimated output of 100 tons of barytes. I have not traced them to any definite locality; but small veins of barytes occur in the schists of the north-east of the county.

CHAPTER IV.

BAUXITE.

Bauxite is now regarded as a mixture of the two aluminium hydroxides, gibbsite and diaspore. Under certain tropical conditions of decomposition, silica is removed in solution from the common complex silicate rocks, and aluminium hydroxide arises in place of ordinary clay. This process of laterisation, as opposed to kaolinisation, gives rise to crusts and often to deep zones of decomposition, in which iron and aluminium hydroxides prevail, together with less altered material from the underlying rocks. Basalt, under these conditions, yields strongly ferruginous material (see under Iron), while lavas rich in silica, such as the rhyolites of Co. Antrim, may produce pale bauxites, in which residual crystals of quartz may still be found. The percentage of silica in a commercial bauxite, in view of the exceptionally good material available from south-eastern France, should be less than 3; and this has militated against the use of much of the Irish bauxite, except in the manufacture of alum. There is a demand, however, for bauxitic clays and bauxite for the manufacture of refractory bricks.

The origin and nature of the Irish deposits has been fully discussed in the Memoir of the Geological Survey on "The interbasaltic rocks (iron ores and bauxites) of north-east Ireland," pp. 1-15 (1912), and the analyses in that Memoir (p. 112 etc.) show a low percentage of silica in several cases, notably in material from Ballynure and Straid (Irish Hill).

BAUXITE

Under the name of Alum Clay, bauxite and bauxitic clays were raised for many years (see Mem. on interbasaltic rocks, p. 60) before works for the production of alumina for the aluminium industry were established at Larne. The discovery of the true nature of the material was made by J. F. W. Hodges, J.P., about 1870, and Kinahan (p. 66) states that the mining of alum-clay began in 1873. It may be noted that an unusual number of misprints unfortunately occurs in this portion of Kinahan's work; few readers, for instance, will recognise the place-name "Margeilleo," attached to an analysis, as "Marseilles," or "Dahm's" as "Dalmis."

The production of bauxite in the United Kingdom was long limited to the county of Antrim, and to occasional supplies from the county of Londonderry. Discoveries in the Carboniferous strata of Scotland may now lead to an extension of the industry. A review in the Report of the Controller of the Department for the development of Mineral Resources, Cd. 9184, p. 23 (1918), shows that records of production in Ireland began in 1882. The output up to 1915 was 292,596 tons, valued at £103,730. The price has shown a tendency to decline from 15s. per ton in 1883 to 4s. 6d. per ton in 1910. Production was to some extent stimulated by the war of 1914-18; in 1915 the output was nearly 12,000 tons, a return towards the figure, 12,402 tons, of 1898. The output from south-eastern France in 1912 was 254,883 tons. The Report points out that the undue amount of silica causes the Irish material to be "mostly used for the making of Aluminium Sulphate required for bleaching purposes in the manufacture of paper."

Kinahan's objection to calling the Irish product bauxite ("Antrim alumyte," Trans. Manchester Geol. Soc., vol. 23, p. 165, 1895) is trivial, and his claim for it of general superiority to the French bauxite implies a misleading of the analyses.

Prominent localities where bauxite has been worked are marked by the aluminum symbol Al on the Mineral Map.

Ballintoy. 1" 7. The pale grey and often pisolitic bauxite of Ballintoy is mined in **Clegnagh** (6" Antrim 4 S.W.), a little south of the high road that leads west from Ballintoy, and there is also an adit on the outcrop nearly 1 mile W.S.W. of Clegnagh, in **Lemnagh More** (6" Antrim 3 S.E.). See Memoir on "Interbasaltic rocks of N.E. Ireland," pp. 22 and 23 (1912). The adit in Clegnagh was closed through lack of timbering when the present writer visited it in 1918, and water had accumulated. The List of Mines for 1918 gives working as suspended. In the mouth of the mine lithomarge can be seen passing up into grey bauxite, and the latter seems to be a bleached product of decay of the basalt, like the whitish material seen in the Giant's Causeway section. Thin slices have revealed pale fragments of scoriaceous lava and also of basalt; both of these may have been derived from underlying basalt. The average thickness at Clegnagh is 2 ft., swelling out to 3 ft. here and there, and at Lemnagh More from 18 ins. up to even 5 ft. in places.

The following analyses have been made for the Geological Survey by Mr. G. Brownlee, in the chemical laboratory of the Department, in 1921, and may be compared with the figures already published on p. 113 of the Interbasaltic Memoir :—

BAUXITE.

	Lemnagh More	Clegnagh
Silica	14·60	18·50
Alumina	33·70	46·48
Ferric oxide	2·40	2·32
Titanium dioxide	10·20	3·70
Calcium oxide	5·62	0·73
Magnesia	0·68	0·29
Potash and Soda	2·39	1·44
Water at 100° C.	3·51	4·40
Loss on ignition	26·34	22·68
	99·44	100·54

Urbalreagh. 1″ 7. 6″ Antrim 6 N.E. See under Iron. This mine is quoted as producing bauxite in the Rep. Controller Depart. Min. Resources, 1918, p. 23.

Killygreen. 1″ 7. 6″ Londonderry 3 S.E. Mem. Interbasaltic Rocks, p. 19. This is returned in Min. Stat. 1914 as an iron mine; but its main produce has been bauxite (see List of Mines for 1915). The adit was closed when the present writer visited it in 1918; but one of the three tips was formed of lustrous brown iron hydroxide. The lithomarge is red and ochreous; the bauxite is brown-grey, and not so pale as that of Clegnagh.

Essathohan. 1″ 14. 6″ Antrim 24 N.E. This mine is one mile north of Parkmore railway station. In Min. Stat. 1898 it is associated with Parkmore and Barard as producing bauxite, and it was worked in 1900 by the Antrim Iron Ore Company. The bauxite of this area is mentioned as being ferruginous in the Memoir on the interbasaltic rocks, p. 54; but the product of Essathohan is not specifically described. The mine is named as raising bauxite in the Home Office List of Mines for 1918.

Cargan and **Evishacrow** are given together in Min. Stat. (1899, for example), as sources of alum-clay, but may now be regarded as iron mines (see under Iron). See Mem. interbas. rocks, pp. 52 and 60, and analysis XLI of bauxitic clay on p. 116, and of bauxite LII and LIII, p. 117. It is of interest to note that the highly aluminous nature of interbasaltic beds in Co. Antrim was first discovered in the Cargan district by J. F. W. Hodges, J.P., about 1870.

Tuftarney. 1″ 14. 6″ Antrim. 24 S.W. Mem. interbas. rocks, p. 50, and analyses XXXIII to XL, p. 116. An outlier of Upper Basalt, forming Tuftarney Hill, between Newtown Crommelin and Cargan, has protected the zone of decomposition, and bauxitic clay has been mined here, resting upon lithomarge. Working in 1918.

Solomon's Drift. 1″ 14. 6″ Antrim 24 S.W. One of the mines in the Skerry Water valley above Newtown Crommelin. See Mem. interbas. rocks, p. 49. Worked about 1910 for bauxite.

Correen. 1″ 20. 6″ Antrim 33 N.W. This mine, just south-east of Knockboy Hill, about 1½ miles north of Broughshane, is given as producing bauxite in Min. Stat. 1898. In the Memoir on interbasaltic rocks, p. 70, a later attempt to reach the bauxite layer is described.

Min. Stat. 1899 give the adjacent **Elginny** mine as a source of alum clay.

Cullinane. 1″ 20. 6″ Antrim 29 N.E. A mine at the hamlet of Cullinane, on an outcrop shown between two faults in the north of 1″ 20, and 2¼ miles S.S.W. of Carnlough, is said to have yielded bauxite extensively prior to 1886 (Mem. 20, p. 16, and Mem. interbas. rocks, p. 76). A plan in the Home Office, No. 3407, records the abandonment of the mine in 1893.

Libbert Mine, Glenarm. 1″ 20. 6″ Antrim 29 N.E. This is marked with a spot south-west of Glenarm on the Mineral Map, and its bauxite bed is well known as having yielded numerous remains of fossil plants when the mine was worked in the closing years of the nineteenth century (Mem. interbas. rocks, p. 30). The whole of the material is probably derived from rhyolite.

On the MS. 6″ Antrim 40 N.W. sheet in the office of the Geological Survey, bauxite, like that of the Libbert Mine, is noted along the outcrop in Old Freehold townland, 2 miles north-west of Kilwaughter village.

Irish Hill and **Straid.** 1″ 28. 6″ Antrim 52 N.W. and 46 S.W. Mem. interbas. rocks, p. 36. The bauxite raised, in common with iron ore, from beneath this outlier of Upper Basalt seems, like that of the Libbert Mine, to be derived from rhyolite.

In Min. Stat. 1881 the whole of the year's output of bauxite, 7,732 tons, valued at £6,765, is attributed to Irish Hill and Straid. The mine is given as raising bauxite only in the List of Mines for 1918.

CHAPTER V.

COPPER.

The mining of copper ore in Ireland, as in Britain, has undergone violent fluctuations, owing to the development of new sources in the Colonies and the United States, and to the variations in demand. The demand, however, has of late largely increased, as a consequence of the spread of electrical industries, and no large body of ore is likely to remain neglected. The Irish mines seem to have worked prosperously in the middle of the nineteenth century whenever copper was near £100 a ton; but we find that £90 a ton was regarded as a low price in 1875 (M.C.I. Report). In 1887 no copper ore was raised in the whole of Munster. The zealous working of the orebodies that were immediately traceable is responsible for the closing of many of the mines as exhausted. A district set with abandoned mine-shafts and mine-cottages presents a melancholy spectacle; but in many cases it represents the foundation of large fortunes and an epoch of industrial efficiency scientifically applied.

The ore of commercial importance in all the Irish mines is copper pyrites (chalcopyrite), crystals of which contain 33 per cent of copper. The " grey ore " of older writers is tetrahedrite, but Kane uses the term for chalcosine. In either case, its occurrence increases the copper-percentage in the ore.

Geevraun (Horse Island). 1″ 40. 6″ Mayo 5 N.E. This lode, north-west of Belderg, is said to have been worked in 1861; it is marked on Griffith's Map (1855), but is indicated by him as unworked (1861, p. 149). It must not be confused with the well-known Horse Island copper mine in 1″ 199. The lode is shown on the 1″ Geological Survey map, running N.E. into Horse Island. There seems to be no recorded output. Reports on Geevraun and the Belderg (*Belderrig*) lode were made by Mr. H. J. Daly to the Department for the development of Mineral Resources in 1917.

A MS. description, with sections of the workings, by Lieut. T. J. Godfrey, R.E., was lodged in the Geological Survey Office, Dublin, in 1918.

Pollboy. 1″ 43. 6″ Leitrim 11 N.W. (N.E. corner). A copper mine was opened in this townland where the road from Manorhamilton to Drumahaire and Sligo turns off from that passing through Glencar. This is evidently the locality where McParlan (Stat. Surv. Leitrim, p. 14, 1802) says that Mr. Wynne raised copper ore from deep pits on the north side of " Binbo." Two lodes, with a north-easterly trend, are shown on the 1" geological map, one being in the Lower Carboniferous dolomitic limestone, and the other in the adjacent Ox Mountain gneiss to the south-east. Griffith knew of the mine (Map and 1861, p. 148), and it is mentioned in Mem. 42 and 43, p. 28 (1885). The Memoir gives some notes of output on p. 29; but there seems to be no official record, though the mine is said to have been working about 1842-6. An opening on the south-eastern of the two lodes is attractively styled "Old Gold Mine" on the engraved 6" Ordnance Survey sheet.

Some copper ore was found in the **Twigspark** lead mines (see under Lead), ½ mile to the north. The western of these two mines is on the border of Shanvaus townland, and is possibly the *Shanvans* copper mine of Kinahan (pp. 31 and 91), said by him to have been worked prior to 1845 (see also Mem. 42 and 43, p. 28).

Sralaghy. 1" 52. 6" Mayo 12 S.E. This lode is somewhat difficult of access, at the head of the Glenamoy River, 2 miles off the road from Belderg to Belmullet, and 5 miles south-west of the former town. It is marked on the MS. 6" sheet of the Geological Survey, and a copy of a report by Lieut. T. J. Godfrey, R.E., was lodged in the Survey Office in 1918. This states that the mine was worked for eight years, from about 1892; but no output seems to have been officially recorded. The name occurs in the Home Office List of Mines for 1892. A report was furnished by Mr. H. J. Daly to the Department for the development of Mineral Resources in 1917.

Bolinglanna (Benderg). 1" 73. 6" Mayo 75 N.E. This is a mine in Old Red Sandstone, east of Corraun village, on the N.W. extremity of Clew Bay. It was known to Griffith (1861, p. 149, and Map, where it is correctly spelt as above). Some galena and iron pyrites occurred with the copper pyrites. The 1" geological map has an indication of copper here, and the 6" MS. map shows a lode running from the coast N. 20° W. S. B. Wilkinson (Mem. 62, 73, p. 21, 1879) mentions two shafts, and notes that the ore "is very rich, but apparently it is almost worked out."

A detailed map showing five lodes in the "Curraun Mining District," and a section of the working on No. 5 in Bolinglanna, was prepared by Edmund Spargo for the proprietor in 1868. A printed copy of this, given by Mr. W. McCormack of Dublin, is in the office of the Geological Survey.

Srahmore. 1" 74. 6" Mayo 65 N.E. Griffith (1861, p. 149) associates this name with Bolinglanna as one of his Corraun Mines, and marks a copper mine on his Map. He records also iron pyrites and argentiferous galena.

The site is 1 mile N.N.W. of hill 1784 of the 1" map, in a peat-covered country set with little lakes, and no trace of a lode or workings was detected by the Geological Survey.

Kinahan, p. 32, evidently copies the name from Griffith, but does not give it as that of a worked mine.

Tullydonnell. 1″ 70. 6″ Armagh 31 N.W. The site of this old mine is engraved on the 6″ map, E.S.E. of Crossmaglen. The M.C.I. (1853 i) reported searches for copper here; but the mine was abandoned in 1853 (M.C.I. 1853 ii). Mem. 70, p. 34 (1877), states that the lode was said to be 2 feet wide, the matrix being "quartz and carbonate of lime."

Salterstown. 1″ 82. 6″ Louth 16 S.W. An indication of both lead and copper is shown on the 1″ geological sheet, 3 miles west of Dunany Point. This is on the site of an exploration by the Hibernian Mining Company some time before 1828 (Griffith, "Mines of Leinster," p. 23, 1828; and Mem. 81 and 82, p. 33, 1871). The shaft sunk was north of the old church of Salterstown and was soon abandoned. See also under Lead.

Beauparc. 1″ 91. 6″ Meath 26 N.W. Copper ore was raised here before Weaver's time (Weaver 1819, p. 282) from a lode in Carboniferous Limestone, on the south side of the Boyne valley, between Navan and Drogheda. Griffith (1861, p. 149) mentions lead in addition to copper. The early workings, the shaft of which was very small, were re-opened and enlarged in 1909, and work was done on the lode of quartz and copper pyrites, which proved to be 2 feet wide. I was able to visit the mine on two occasions while it was at work. Min. Stat. 1909 record 4 tons of ore, yielding 13·75 per cent. of copper; 1910, no record; 1911, 330 tons, 15 per cent. copper; 1912, 500 tons, with 10·7 per cent. copper, valued at £4,000; 1913, 167 tons; 1914, mine ceased working. Proposals for re-opening were made in 1918, and the lode is held to extend favourably westward. It was reported on to the Department of Mineral Resources in 1917.

Brownstown and **Cusackstown.** 1″ 91. 6″ Meath 32 N.W. This area, which is, like Beauparc, in the Carboniferous Limestone, lies E.S.E. of Navan, and was viewed hopefully by Griffith ("Mines of Leinster," p. 25, 1828). Stewart (p. 105) says that a trial, producing rich ore, was made "many years" before 1800. Under the name **Walterstown,** R. Thompson (Stat. Surv. Meath, p. 25, 1802) says that miners were at work in 1802, on a vein running N.E. towards the Boyne; in Appendix 7 to his work, p. 33, an analysis of the ore by J. Purcell is quoted, showing 20 per cent. of copper, which, of course, by itself is no guarantee of the richness of the lode. The MS. 6″ Geological Survey map records searches in the south of Cusackstown "commenced as far back as 1795, and carried on at far apart intervals up to 1840." The symbol for copper appears on the 1″ sheet west of Somerville House and also at Cusackstown, which lies to the south-west; but the development contemplated by Griffith was never attained.

Cleggan. 1″ 93. 6″ Galway 22 N.W. A small mine was opened here about 1830, when 60 tons of ore were shipped from Cleggan Bay (Mem. 93 and 94, p. 162, 1878).

Teernakill South. 1″ 94. 6″ Galway 38 N.E. The Min. Stat. Lists of Mines for 1865-75 give *Ternakill* as raising "copper and pyrites," under Henry Hodgson, who worked the lead mines in Glenmalur. The MS. 6″ map of the Geological Survey records, east of the road from Maam Cross station to Maum, in a hummocky, country where granite intrudes into amphibolites and schists, " mining carried on here on pyritous ore in hornblende-schist. Limestone blocks brought up apparently with the ore." Memoir 93, 94, p. 163 (1878), mentions Hodgson's work, and says that some pyrrhotine was found. See under Nickel.

Dooros. 1″ 95. 6″ Galway 39 N.E. This mine is on the long promontory running into Lough Corrib, north of Maum Bay. The M.C.I. (1854 i) held that it showed good copper prospects ; £584 were spent in searches (1855 i), but the work was abandoned in 1855 (1855 ii). Mem. 95, p. 67 (1870), speaks of a shaft put down about 1860, which may possibly be that of the M.C.I.

Glann Mines. 1″ 95. 6″ Galway 39 S.E. and 40 S.W. The district of Glann (so spelt on the 1″ map of 1902 ; " Glan " on early issues) is on the south-west shore of Lough Corrib and north-east of Oughterard. The granitoid and metamorphic rocks are here rich in mineral veins. Subrine, Engineer to the King of France, in " A Memoir on the Mines of Glan, the Royalty of Richard Martin, Esq." (Trans. R.I. Acad., vol. 8, p. 141, 1802), describes the veins as in slate, and criticises the mining proceedings that were then being carried on in search of copper ore (pp. 148-156). Mem. 95, pp. 60-63 (1870), names several mines, in which copper pyrites was often associated with iron pyrites and sometimes with galena. Occasionally pyrrhotine was accessory. Some of the most important trials seem to have been made at the mine of **Curraghduff.** Min. Stat. Lists of Mines cite **Glanmore** as a mine of copper ore and pyrites from 1865-8 (H. Hodgson proprietor). See under Sulphur.

Claremount. 1″ 95. 6″ Galway 54 N.W. This mine, described in Mem. 95, p. 64 (1870), is situated near the high road just west of Oughterard, and is primarily one of the lead mines of this district. It yielded, however, some copper pyrites. The shaft was sunk in granite.

Truska. 1″ 103. 6″ Galway 49 S.E. and 50 S.W. The mine was opened on a lode south-east of Truska Lough, a little north of Dog's Bay and west of Roundstone (there is also a townland of this name south-west of Clifden). The mine appears to be a modern venture. Min. Stat. 1891 record 4 tons of ore as the output.

Errisbeg (*Errisberg*). 1" 103. 6" Galway 63 N.W. This mine, in Errisbeg West, on Dog's Bay, west of Roundstone, was worked by the Errisbeg Mining Company under M. G. Painter, from 1880 to 1887. It appears as *Errisbey* in the List of Mines in Min. Stat. 1878-9. Twenty tons of ore, yielding 1 ton 5 cwt. of copper, were raised in 1880 (Min. Stat.), and 40 tons, realising £145, in 1881. In 1886 this was the only mine raising copper ore in Ireland, since the Wicklow mines were then depending on precipitates. The name is quoted in 1887, but no output is given for 1886 or subsequent years. The 6" MS. map of the Geological Survey records a trial shaft, and a quartz vein containing copper carbonate on the coast.

Loughshinny. 1" 102. 6" Dublin 8 N.E. The excavations in the contorted Carboniferous rocks of this small sea-inlet are well known to residents in the county, but no success seems to have attended the mining operations made in various years. John Rutty ("An Essay towards a natural history of the county of Dublin, vol. 2, p. 142, 1772) records that Benedict Arthur, of Seafield, "raised some Copper Ore, but dropt it, the Vein, tho' sometimes affording large lumps, proving thready and small." Stewart (p. 68) mentions that English miners were employed to bore here for coal about 1775, and he hints that it would not be to their interest to report favourably!

Workings for copper were renewed in 1807-8, and Hunt (1848) gives small outputs of ore for 1808-9. Weaver (1819, p. 237) mentions an apparently rich ore in quartz, which was not being mined when he wrote. Griffith ("Mines of Leinster," pp. 23 and 43, 1828) describes the workings as extensive "within the last thirty years." These were abandoned, and then taken up again with no more success by the Hibernian Mining Company. Hunt (1848) records 18 tons of ore as raised in 1844. Du Noyer (Mem. 102 and 112, p. 70, 1875) says that the mine had "long ceased to be worked" in 1849. It was probably closed, therefore, in 1845. The Memoir gives some description of the lodes.

Castleknock. 1" 111. 6" Dublin 17 N.E. Rutty ("Nat. Hist. Dublin," vol. 2, p. 142) mentions copper ore as having been raised before 1772 in Castleknock parish, on T. Kennan's estate, near Diswellstown House, a mile west of Knockmaroon Hill. Mining continued for only a few months, and the ore did not defray expenses. In view, however, of possible future discoveries in this limestone region, the matter seems worth recording.

The Ovoca (Avoca) Mineral Belt.

This important mineral area, which will be referred to also under ores of iron and sulphur, was primarily opened up for copper-mining. The mines were developed before the middle of the eighteenth century on the flanks of the steep and picturesque valley of the Avonmore, which, below the entry of its important tributary, the Avonbeg, is known as the Avoca River. The Ordnance Survey, down to 1905, wrote both the village and river names as Ovoca, following the "Oboka" used by Claudius Ptolemaeus about 150 A.D. (Joyce, "Irish Names of Places," p. 75, 1871); but in the 6" map of 1910 the form Avoca was adopted for both the village and the river. The spelling Ovoca is retained in the present Memoir, owing to its very general use in literature dealing with the mines. At Woodenbridge, another of the consequent streams from the Leinster Chain, the Aughrim River, known also as the Daragh or Darragh, comes in from the north-west, and the mines farther to the south are styled collectively the **Carysfort Mines** (Kinahan, p. 118).

The deposits are approximately in the strike of the Ordovician strata of the east flank of the Leinster Chain, and are, no doubt, connected in origin with the igneous rocks, ranging from altered rhyolites to dolerites, that form so conspicuous a feature in the county of Wicklow. They have been aptly compared with those of Huelva in Spain by A. M. Finlayson ("Petrology of Huelva," Geol. Mag., 1910, p. 228, and "Metallogeny of the British Isles," Quart. Journ. Geol. Soc. London, vol. 66, p. 284, 1910). The copper ore is in large part cupriferous iron pyrites, but considerable quantities of copper pyrites were extracted during the early working of the district. A possible relation of the pyrites ores, whether of copper or iron, to bodies of magnetite is shown by the occurrence of magnetite lodes in the same belt, and with the same prevalent strike, at both ends of the mineral region, namely at Ballard and at Moneyteige.

The smelting was done at Swansea, though works were erected early on the Arklow shore (Griffith, "Mines of Leinster," p. 20, 1828; see also under Ballymurtagh); these were abandoned owing to the impossibility of landing coal during the winter months. Most of the mines have carried on a considerable trade in copper precipitated by iron from the mine-waters. (Wm. Henry, "Copper Springs in the County of Wicklow," Phil. Trans., vol. 47, p. 502, 1753; Weaver 1819, p. 218; Smyth, "Mines of Wicklow," p. 385; and Mineral Statistics for successive years.)

The development of the district for the production of sulphur ore rather than copper ore (see under Sulphur) took place in 1840. Griffith does not mention sulphur as important in 1828, though a certain amount had already been extracted from the copper ore on the spot (Weaver, 1819, p. 218). In Smyth's memoir (1853), on the other hand, sulphur appears as all-important. During the European War of 1914-18, the small percentage of copper in some of the "sulphur-ores" was again commercially extracted.

This productive mining region has given rise to a considerable literature. R. Fraser (Stat. Surv. Wicklow, p. 15, 1801) deals with the important mines of Ballymurtagh and Cronebane. An excellent account of the mines at the end of the eighteenth century is given by Weaver (1819, pp. 213-219), who was resident acting partner of the Associated Irish Mine Company. Kane (1845, p. 184) brings Weaver's description up to date. The Geological Survey, under T. Oldham, issued a special sheet of plans and sections of the Ovoca mines about 1852, and more recent workings were inserted on the second edition, published in 1879. The two editions can be distinguished at a glance from the fact that on the earlier one the igneous rocks are coloured green and on the latter red.

It was left to Warington W. Smyth, as Mining Geologist to the Geological Survey of the United Kingdom, to report officially on "The Mines of Wicklow and Wexford" (Records of the School of Mines, vol. 1, part 3, 1853), and his admirable pamphlet, with maps and illustrations, is fortunately still procurable. J. Beete Jukes proposed at that date to write a memoir on the geology of the district (Smyth, p. 409); the 1" maps 130 and 139 were published in 1855 and 1856 respectively; but the memoir on the former was delayed until 1869, and the description of the principal mining area (Mem. 138 and 139, by Hull, Cruise and Hatch) did not appear till 1888. A short account of the district was given in 1860 by F. W. Egan and H. Geoghegan (G.S.D., vol. 8, p. 179). P. H. Argall, who studied the ground with G. H. Kinahan of the Geological Survey, provided new plans and sections of the eastern Ovoca mines, and a general description of the lodes and workings, in his "Notes on the ancient and recent mining operations in the East Ovoca district" (R.G.S.I., vol. 5, p. 150, 1880). Kinahan ("Economic Geology of Ireland," pp. 109 and 480, 1889) has summarised his own wide knowledge of the district. The mines were reported on in 1917 to the Department for the development of Mineral Resources (Report of Controller, 1918, p. 54). A list of some of the papers on the Ovoca district is given in Memoir 138 and 139, p. 37 (1888). In the following notes the copper mines are arranged from north to south, that is, in order down the valley. From Kilmacoo to Tigroney they lie on the east bank, and from Kilcashel to Ballymurtagh on the west. A large amount of statistical matter has been examined, and certain salient features as to working and output have been selected here for mention. Though copper and small quantities of gold have been regularly extracted from the ores, the interest of Ovoca in recent times has lain in the production of sulphur and ochre. Hence the principal mines will be again referred to under the head of Iron and Sulphur, while some also appear as sources of Zinc.

During the last four years the Electrolytic Copper Company has been systematically carrying out extensive boring and exploratory work in the Ovoca valley, and very large bodies of low grade copper ore have been proved as the result of their investigations. Hence there is good reason to believe that the district will become highly important for the production of copper, as well as of lead and zinc. There should, of course, be in addition an increased production of sulphuric acid (see Report of the Controller, Depart. development of Min. Resources, p. 40, 1918).

MS. transverse sections of the Ovoca mines, by G. H. Kinahan, are in the Geological Survey Office, Dublin.

Kilmacoo. 1" 130. 6" Wicklow 35 N.E. This mine lies in Kilmacoo townland, high above the Ovoca valley, and is known for its "bluestone" (Weaver 1819, p. 215; Argall, R.G.S.I., vol. 5, pp. 160 and 164, 1880). This is a mixed galenitic ore, known also as "kilmacooite"; it will be again referred to under Zinc.

Connary (*Connaree, Connery, Connarree* and *Connoree*). 1" 130. 6" Wicklow 35 N.W. Mem. 121 and 130, p. 44 (1869). From 1832-45 (Hunt 1848, p. 713) 11,380 tons of ore were raised from this mine. While the general yield in metallic copper of the Ovoca ores runs from 2 to 8 per cent., Min. Stat. 1852 quote from Connary 27 per cent.; 1853, 27⅝ per cent.; 1860, for 14 tons of ore, 37⅞ per cent.; 1862, for 51 tons of ore, 38¾ per cent. This seems to indicate that ores other than copper pyrites were occasionally met with. In 1863, 3,682 tons of ore yielded the more normal amount of 5⅛ per cent. of copper; in 1865 only 79 tons of ore were raised, bearing 8¾ per cent. The decline in copper mining in the British Isles must have affected Connary, which became noteworthy as a producer of sulphur ore (Mem. 121 and 130, p. 44, 1869). No return of copper from it appears in 1884. Min. Stat. 1858 quote returns from precipitates obtained at this mine from 1856-8. Smyth (p. 387, 1853) states that the precipitate here gave only 43 to 54 per cent. of copper, owing to a somewhat rapid process of collection.

A section of "Connorree," showing levels down to 54 fathoms, with comparatively small portions marked as "worked out," was deposited (R 190 and 414) in 1876 in the Home Office as that of an abandoned mine.

See also under Lead, Sulphur and Zinc.

Cronebane. 1" 139. 6" Wicklow 35 N.W. and N.E. The eastern mine of this name is also known as the **Magpie**, from its being worked on the Magpie lode, and is associated with Connary. **West Cronebane** lies to the south-west, and its produce is often tabulated with that of Tigroney (for descriptions see Smyth, p. 389; Argall, R.G.S.I., vol. 5, p. 156; Kin., p. 113). Cronebane is stated by W. Henry (Phil. Trans., vol. 47, p. 501, 1753) to have employed 500 men at eightpence day as far back as 1752; and Griffith ("Mines of Leinster,"

p. 19, 1828) says that it was discovered and effectively worked "more than a century ago," say, 1720. After a period of neglect, one of the owners, Mr. Mills, revived the mine in 1787. Through the incorporation of the Associated Irish Mine Company in 1798, T. Weaver, as above mentioned, became resident acting partner at Cronebane (Weaver, 1819, p. 214). The ore raised from 1787 to 1811 gave about $6\frac{1}{2}$ per cent. of copper (*ibid.*, p. 217). From 1823-47 (Mem. Geol. Surv. Gt. Brit., vol. 2, pt. 2, p. 713) nearly 38,000 tons were raised, culminating with 3,112 tons in 1836. The figures for the percentage of copper are high for certain years ($40\frac{3}{4}$ per cent. in 1850 for 13 tons of ore); but Smyth states that "the sulphuret and black oxide of copper" formed the principal deposit. In 1866 (Min. Stat.) only 62 tons of ore are quoted. The mine then drops out of official reports, but reappears in 1875 with 847 tons of ore. There is no record from 1881 to 1897, after which the output, given as "cupriferous pyrites," is combined with that of Tigroney until 1912. In recent years, the copper content of the ore has been about 1·2 per cent.

W. Henry (*op. cit.*, p. 501) states that in 1752 "Crone Bawn" was worked by several shafts 50 to 70 fathoms deep. The first ore met with on the hill-top was "iron stone," succeeded by silver-lead ore, and then a "very rich" copper ore. The extraction of copper by placing iron bars in the mine-waters was discovered in Cronebane through the accidental leaving of a shovel in the outflow, not long before 1750 (*ibid.*, p. 502). Rutty refers to this ("Nat. Hist., Dublin," vol. 2, p. 87, 1772). A plan and section of Weaver's Boat level, Cronebane, by P. H. Argall, are in the Office of the Geological Survey, Dublin. Weaver (1819, plate 12) gives an early plan and a section of the mine.

Tigroney (*Tigrony*). 1″ 139. 6″ Wicklow 35 S.W. The working of this mine has been associated with that of West Cronebane (Argall, R.G.S.I., vol. 5, p. 151). From 1822-47 (Mem. Geol. Surv. Gt. Brit., vol. 2, part 2, p. 713), 24,590 tons of ore were raised from Tigroney, with a maximum output of 1,800 tons in 1837. As in the case of Cronebane, small quantities of abnormally rich ore are quoted about 1854. The ordinary output has been cupriferous pyrites with about 3 per cent. of copper, and also precipitate with 50 to 60 per cent. (Smyth, p. 387). Weaver (1819, plate 12) gives an early plan and a section of the mine.

Knockanode. 1″ 139. 6″ Wicklow 35 N.W. This name is not indicated by a spot on the Mineral Map. It was opened on the west bank of the river just below the Meeting of the Waters, probably in the hope of finding lead (Weaver, 1819, p. 219; Kin., p. 483). Plans have been deposited with the Home Office as those of an abandoned copper mine (R 284).

Kilcashel. 1″ 139. 6″ Wicklow 35 S.W. This townland lies immediately south of Knockanode. Weaver (1819, p. 219) records trials here for copper ore. Griffith (1861, p. 154) records the working of copper and sulphur ore here by the Wicklow Copper Mine Company. "Whaley's Shaft," marked on the published "Plan and Sections of the Ovoca Mines," is in the south of Kilcashel.

Ballygahan. 1″ 139. 6″ Wicklow 35 S.W. Kin., p. 116. Smyth, "Mines of Wicklow," p. 376. This mine is in the townland of Ballygahan Lower, which occupies the river-bank east of Ballymurtagh. It is about a mile north-east of the word indicating Ballygahan townland on the map in Smyth's Memoir, and the excavations are actually north-east of those of Ballymurtagh. 3,801 tons of ore were raised here from 1828-47 (Mem. Geol. Surv. Gt. Brit., vol. 2, pt. 2, p. 713). About 1855 the copper percentage was 4¼. The output rose from 651 tons of ore in 1863 to 2,357 tons in 1864; by 1879 a marked decline had set in, and not even a precipitate trade was carried on from this or any other Wicklow mine in 1892. Sulphur ore, however, rose greatly in importance in the interval.

Ballygahan is quoted in Min. Stat. Lists of Mines as worked for copper and pyrites by Henry Hodgson from 1860 to 1877.

Ballymurtagh. 1″ 139. 6″ Wicklow 35 S.W. This mine has been of great importance in the district as a source of copper, sulphur and ochre. William Henry (Phil. Trans. vol. 47, p. 500) refers to it as disused in 1752, owing to " some difference between Mr. Whalley [Whaley] and the Company," although it was previously very profitable. Holdsworth ("Geol. etc. of Ireland," p. 24, 1857) speaks of it as "very successfully wrought for copper" in 1755. It was certainly re-opened by Whaley about 1768, and the proprietors have the credit of erecting the Arklow smelting works already referred to (Fraser, "Stat. Surv. Wicklow," pp. 15 and 16, 1801). From 1822-47, the annual output was very variable (Hunt 1848); but the total copper ore raised in these twenty-five years was 52,111 tons. The copper-content from 1853 to 1868 is given in Min. Stat. as near 4 and sometimes 5 per cent. The return for 1874 shows 8 per cent.

The history of Ballymurtagh is given by S. Haughton ("Geol. and Stat. notes on Irish mines," G.S.D., vol. 5, p. 279, 1853), with a description of the lodes and (p. 282) a section of the mine. At this time (Smyth, "Mines of Wicklow," p. 374, 1853) the mine gave employment to 800 persons. Serious falling-in of the ground has occurred here, notably in 1835 and 1845, as a consequence of the extensive excavations. From 1862 to 1879 Ballymurtagh was worked by the Wicklow Copper Mining (or Mine) Company. The great development about 1867 will be recorded under Iron and Sulphur. Recently its output has been ochre. It was reported on by Mr. H. J. Daly to the Department for the development of Mineral Resources in 1917.

A plan of Ballymurtagh in 1845 appears in the Home Office series, though the mine does not seem to have been abandoned at that date.

Knocknamohill. 1" 139. 6" Wicklow 40 N.W. The site of this mine is given under Iron. Griffith (1861, p. 164) mentions the raising of copper here, and the date of this appears to have been about 1846. There is no official record. Kinahan (p. 34) mentions copper mining here, and at the adjacent mine of **Ballymoneen** (see Iron). In the N.E. part of Knocknamohill "ancient open works" and "copper mines here about 1846 (Crockford and Company)," are marked on the 6" MS. map of the Geological Survey.

Moneyteige. 1" 139. 6" Wicklow 39 S.E. The magnetite lode in Moneyteige South includes some copper ore; it was worked as one of the Carysfort mines (Kin., p. 118). See under Iron.

Ballyvergin (*Ballyvirgin*). 1" 133. 6" Clare 26 S.E. This mine is better known for its production of lead ore; but it appears in Min. Stat. 1856 with 275 tons of copper ore, yielding some 8 per cent. of copper. Ninety-one tons of ore were raised in 1857; but in 1858 it is recorded as a lead mine.

The Silvermines District.

While this mineral district has become known since 1858 mainly through the ores of zinc, it produced a certain amount of copper in former days. The two copper mines are marked on the 6" Ordnance map; they lie on the great line of fracture that has lowered the Carboniferous strata on the north, and has allowed of the remarkable mineral impregnation of the district.

Gorteenadiha (*Gurteenadya, Gurtnadyne, Gortnadine*). 1" 134. 6" Tipperary 26 S.W. Memoir 134, pp. 34 and 44, (1861). This mine is on the north side of the great fault, in Lower Carboniferous shale, and just east of the stream in Gowlaun Glen. It is indicated by a symbol on the 1" Geological map. The gangue of the lode is a quartz-breccia with some barytes, a mineral common along the line of fracture. Lead ore was raised here as well as copper pyrites, and Griffith (1861, p. 152) states that both were argentiferous. Gorteenadiha was worked in the eighteenth century, and for a time about 1801 (Weaver 1819, p. 242). Min. Stat. 1850 give the output as 154 tons of ore, with $10\frac{5}{8}$ per cent. of copper.

In Min. Stat. 1857 the mine is placed in County Wicklow with 130 tons of ore ; the output sank to 42 tons in 1861. In Min. Stat. 1861 " Gurtnadyne " appears also in County Waterford, as a lead mine. In the Lists of Mines for 1860 and 1861 (p. 25), it is given (again under Waterford) as producing lead and pyrites, not copper. Copper is again mentioned, in addition to lead, from 1865-71. As a lead mine, Gorteenadiha was worked until 1870, and it remains on the list till 1874.

Ballynoe (Gortshaneroe, *Ballinoe*). 1" 134 (name engraved on geological map). 6" Tipperary 26 S.E. Memoir 134, pp. 37 and 44, 1861. This mine was near the centre of the western boundary of Gortshaneroe (*Gortshanroe*) townland, less than a mile S.W. of Silvermines. The pit was sunk 80 feet, and a great bunch of ore was taken out of dolomite rock. It yielded small quantities of copper ore in 1850 and 1851, and in 1858 to 1860 (104 tons in 1859). Min. Stat. 1860 place it in County Wicklow. It was also known as a lead mine from 1865-9. A. Gages (G.S.D., vol. 8, p. 243, 1860, and A. B. Wynne, *ibid.*, p. 246) described orpiment in barytes at Ballynoe.

Coolruntha (Goldmines). 1" 134. 6" Tipperary 32 N.W. Two crossing lodes are shown on the Geological Survey map at the hamlet of Goldmines, which lies on the western margin of the townland of Coolruntha, at a height of 800 feet above the sea on the slope of Slievekimalta. Though no output has been recorded, Mem. 134, p. 42, states that the traces of old workings are considerable.

Ballyhourigan. 1" 134. 6" Tipperary 32 N.W. A mile to the south-west of Coolruntha mine, on the northern border of Ballyhourigan wood, the 6" MS. map of the Geological Survey records a shaft. This is in the south-west of Barnabaun townland. A. B. Wynne, in 1861 (Mem. 134, p. 42), could obtain no evidence in regard to it, though local information, as usual, reported that the copper ore was " rich." Wynne learnt later (Mem. 133, p. 36, 1862) that there were two shafts, 40 and 35 feet deep, respectively, and that the copper vein was 4 inches wide. Mr. H. J. Daly reported on the mine to the Department for the development of Mineral Resources in 1917.

Rathnaveoge. 1" 135. 6" Tipperary 17 S.W. This old mine is some 6 miles S.W. of Roscrea ; Griffith, 1861, p. 151, gives it, from his MS. notes, as perhaps worked in the seventeenth century. A. B. Wynne (Mem. 135, p. 32, 1860) records a lode, traced chiefly by indications of old shafts, bearing about E. 10° N.

Pallaskenry. 1" 143. 6" Limerick 3 S.E. Mem. 143,

p. 34 (1860) records tetrahedrite and bornite in fragments of quartz gangue; the sinking, though it had been recently made, was closed when visited in 1856. The mine was in the more northern of the two Ballydoole townlands; Kinahan (pp. 20 and 31) implies that some lead ore was at one time raised from it. An older mine was found when sinking a pump-hole at the Old Charter School-house, a little to the N.W. (Mem. 143, p. 34).

Lackamore. 1" 144. 6" Tipperary 38 N.W. The mine is on the southern border of the county, on a lode in Silurian shales, near the passage of these strata under the Old Red Sandstone that forms the hill-mass above Newport. References to it occur in Weaver 1819, p. 243; Kane 1844, p. 189; and A. B. Wynne in Mem. 144, p. 36, 1860, who describes the conditions in 1859. Kinahan, p. 101, says that "the Mining Company" worked here some time after 1810; but there is no reference to the mine in the reports of the Mining Company of Ireland. Min. Stat. 1858, p. 38, quote 37 tons of argentiferous lead ore as raised from Lackamore; but the mine is placed in Clare, and the record is probably an error.

Weaver describes "two veins composed of brown spar, calcareous spar, clay, and iron ochre, more or less indurated, a few inches in width, and a third vein, composed of the same materials, but of greater thickness, bearing rich copper ore in bunches." Workings were made (about 1810) 120 fathoms long and 26 fathoms deep, with an engine-shaft reaching 36 fathoms. The pumping arrangements proved insufficient to free this low-lying mine from water. Shortly before 1844, however, the works were reopened (Kane 1845, p. 199), and 200 people were employed. Min. Stat. give figures from 1837 to 1859, showing 648 tons of ore raised in 1841, but nearer 100 tons annually in subsequent years. Messrs. John Taylor and Son, of London, were the last workers, and closed the mine in 1859. Though the ore was said in 1848 to yield 22 per cent. copper, the average proved to be about 9 per cent. The mine is recorded in Min. Stat. Lists of Mines down to 1865; from 1862-5, Miss Hamilton is given as the owner.

A report on Lackamore was made by Mr. H. J. Daly to the Department for the development of Mineral Resources in 1917.

Killeen. 1" 144. 6" Tipperary 38 N.W. This mine is in Killeen townland, north of Lackamore, and has been called Killeen Lackamore. It must not be confused with Killeen North mine, known also as Kilcrohane, in County Cork (Kin., p. 27). It was worked during the revival of copper mining from 1905 to 1909, the ore being rich (19–25·7 per cent. copper). Min. Stat. give 35 tons of ore for 1905; but a report by Mr. Sydney Smith to Mr. C. C. Singleton records 116 tons 12 cwt. from September 1905 to September 1906. The next and final record in Min. Stat. is 31 tons in 1909. Mr. Daly reported on the mine to the Department for the development of Mineral Resources in 1917.

Hollyford. 1″ 145. 6″ Tipperary 45 N.E. The name was spelt *Holyford* on the older 1″ Ordnance map. Kinahan calls the eastern mine **Ballycohen** on p. 32, and *Ballycolein* on p. 101; the former name is correct. See Memoir 145, p. 33 (1860), for a good account of the lodes by A. B. Wynne, with plan and section by J. Pascoe (1858), and Geological Survey Longitudinal Section 10, section 6 (1861). The M.C.I. took up the mines in 1837. Troubles with water caused a suspension in 1839, and the leases at " Hollyford and Ballisanode " were cancelled in 1840. The Min. Stat. show that the mines were again worked in 1848 to 1862, the ore averaging nearly 18 per cent. copper. 555 tons raised in 1854 were valued at £10,919. In 1862, however, only 10 tons of ore were raised, and the record ceases. "Holyford" appears, however, in Min. Stat. Lists of Mines 1865-8 as held by the Holyford Mining Company. Mr. H. J. Daly reported on these mines to the Department of Mineral Resources in 1917.

Oola Mines (Oolahills). 1″ 154. 6″ Limerick 24 S.E. and 25 S.W. Memoir 154, p. 28 (1861), and Griffith 1861, p. 148. Some copper pyrites was raised from these mines when worked about 1855, and the name appears as that of a copper mine in Min. Stat. List of Mines, 1865. See under Lead.

Ballynakill. Co. Kilkenny. Tighe (Stat. Surv. Kilkenny, p. 30, 1802) mentions that shafts were put down in an unsuccesssful search for copper at Ballynakill, on the bank of the Nore between New Ross and Inistioge, "many years ago." It seems probable that the site was Ballynakill townland (1″ 157. 6″ Kilkenny 29 S.W.), 5 miles N.E. of Inistioge. Ordovician slates occur here comparable to those associated with the Wicklow mines, and the adventurers are said to have been Wicklow men. The sinkings are mentioned here, since local tradition may some day raise a question of the Ballynakill exploration.

Knockatrellane (Ballymacarbry). 1′ 166. 6″ Waterford 5 S.E. This mine lies in the passage between the Old Red Sandstone masses of the Knockmealdown and Monavullagh Mountains, south of Ballymacarbry village (spelt also Ballymacarbery and Ballymacarberry). Griffith mentions it from his MS. notes, but no details seem available (1861, p. 152.) Du Noyer (Mem. 166, p. 9, 1858) says that no true lode existed, the ore being merely disseminated in the sandstones.

Ross Island. 1″ 173. 6″ Kerry 66 S.E. There is only a casual mention of this mine in Mem. 173 (1861). The history of the working of the lode, which occurs in flinty Carboniferous limestone, is, however, given in much detail by Weaver (1838, p. 30), who provides a plan and section.

The first excavations were ancient. In the six years following the reopening in 1804, 3,220 tons of ore were raised. Weaver took up operations from 1827-9, the output being 1,529½ tons of ore, averaging 13⅔ per cent. of copper. Mining then seems to have been abandoned. Kane (1845, p. 198) states that this was due to flooding by the lake waters; but it is also likely that no great extension of the ore, either laterally or in depth, was to be expected. The ore was not in definite lodes, but was irregularly disseminated in the Carboniferous Limestone. A fine coloured map, showing numerous shafts, and also a section, made by Weaver in 1829, are in the Home Office series (R 286).

Crow Island. 1" 173. 6" Kerry 66 S.E. Weaver (1838, p. 34) mentions about 100 tons of copper ore as having been raised from a longitudinal excavation some 6 fathoms deep on Crow Island, half a mile east of Ross Island. Hunt (1848) records a total sale at Swansea of 86 tons of ore in 1812 and 1813. No further record occurs.

Muckross (*Muckruss*). 1" 184. 6" Kerry 74 N.E. This mine is so near Ross Island that it is convenient to consider it before proceeding eastward. A notice of it by — Wright was read before the Geological Society of London in 1809 (Trans. Geol. Soc. London, vol. 5, p. 595, 1821). It was then stated that the lode of "yellow copper ore" was worked, near traces of far older shafts, from 1749-1754, to great advantage. "Dark blue" cobalt ore (probably smaltine) "tending to a beautiful pink" (erythrine) was found, but was landed as rubbish; one adventurer, however, took away 20 tons. Kinahan (p. 86) states that this ore was discovered by Raspe in 1794. R. E. Raspe, by the by, after a career as a mining adventurer, died at Muckross in that year. He will be best remembered as the author of "Baron Munchausen's Narrative of his marvellous Travels," published anonymously in London in 1785. Weaver (1838, p. 29) gives a plan and section of the mine, and says that the arsenical cobalt ore was a band half an inch to two inches wide, "more or less intermingled with the copper ores." He states that, after the first period of working, mining was resumed on the eastern side in 1785 and 1801, and on the western side in 1795 and 1818, the depth here being about 30 fathoms. The ore at one time (according to Smith, "History of Kerry") gave 26⅔% of copper. Though Du Noyer (Mem. 184, p. 37, 1859) says that the ore occurred in a true lode, Weaver (1838, p. 29) regarded it as layers and strings disseminated in slate and limestone, and the abandonment as due to the ends of the workings becoming narrower and less productive.

The Knockmahon Mining Area.

This area, on the southern coast of the county of Waterford, is included in 1" sheet 178, a revised edition of which was issued by the Geological Survey in 1901, and again, with small alterations in colouring, in 1913. The lodes, which contain some lead ore as well as copper ore, occur in the complex series of Ordovician (Lower Silurian) sediments, volcanic rocks, and intrusive masses, all earlier than the Upper Old Red Sandstone, which appear in the cliff-sections and strike north-eastward along the west flank of the Leinster Chain. The exposure of the lodes on the sea-shore probably attracted attention in prehistoric days. J. H. Holdsworth read a paper on the geology of the district in 1833 (G.S.D., vol. 1, p. 85, 1838), in which he mentions the recent development of the copper mines. In the joint Memoir to sheets 167, 168, 178 and 179 (1865), G. V. Du Noyer briefly describes the lodes and the previous history of mining in the area. Since then, the geological details have been closely studied by F. Cowper Reed (Quart. Journ. Geol. Soc. London, vol. 53, p. 269, 1897, the red rocks of Old Red Sandstone age; vol. 55, p. 718, 1899, Lower Palaeozoic sediments and volcanic rocks; vol. 56, p. 657, 1900, volcanic and intrusive rocks). In these three papers Knockmahon is referred to on pp. 281, 737 and 667, etc.

Systematic mining began on the Knockmahon area in 1730; but Weaver (1819, p. 248) merely mentions copper mines near "Bonmahon," as "formerly worked to some extent." Stewart (p. 134) says that a mine at Bunmahon was working with profit in 1800. The real development came when the M.C.I. acquired rights at **Knockmahon** (6" Waterford 25 S.W.) in 1824. The Reports of the Company and the Min. Stat. have supplied most of the details given below. The table on p. 713 of the Mem. Geol. Surv., Gt. Britain, vol. 2, part 2 (1848), gives figures of output from 1826 to 1847. The M.C.I. made trials at Knockmahon, sank shafts, and sent a small parcel of ore to Swansea in 1826, which yielded $12\frac{5}{8}$ per cent. of copper. It appears from the Geol. Surv. table that **Kilduane** mine (6" 25 S.W.), also known as the **Glen** mine, half a mile to the N.N.E. of Knockmahon, was opened in 1828, but was then for a time abandoned, being again worked in 1845, when the M.C.I. Report states that its shaft was 110 fathoms deep. In 1829 the M.C.I. took up leases at **Bunmahon** (then commonly spelt *Bonmahon*, 6" 24 S.E.) farther west, raising 100 tons of ore in 1830. Bunmahon was evidently at work in 1844, when T. Oldham made his observations on temperature in mines (G.S.D., vol. 3, p. 72, 1849). The **Moneyhoe** mine (6" 24 S.E.) was opened hard by, on some old workings, in 1836. The name is not that of a townland, and Weaver (1838, p. 25) writes it as *Monachoe*. Weaver, who prepared his paper in 1835, had no great opinion of the district, being perhaps misled by the number of old abandoned trial-holes. Searches continued at Bunmahon until 1845, but nothing profitable was found. During these years, however, Knockmahon was progressing very rapidly, the price of £100 per ton for copper in 1840 allowing a profit on the year of £16,728. Smyth's Home Office plan (see later) is dated 1845. All the smelting was done in Swansea. A section of the "Stage Mine," Knockmahon, copied in 1845, is in the office of the Geological Survey in Dublin. The Knockmahon shaft reached 180 fathoms in 1847.

Ballynasissala (*Ballynasisla* of Weaver 1838, p. 25), north of Knockmahon, was also opened ; but, owing to a fall in prices and to high royalties, both this mine and Knockmahon were actually working at a loss in 1848 and 1849. An Act of Parliament then gave relief to mining leaseholders. A MS. section of Ballynasissala, as made for the M.C.I., is in the Geological Survey Office in Dublin.

Kilmurrin, 1½ miles N.E. of Knockmahon, was proved by a shaft in 1849, and gave favourable indications up to 1851 ; the lode was, however, abandoned before 1865.

The **Tankardstown** lode (6" 25 S.W.), a little to the east of the Knockmahon lode, was worked in 1850, and proved one of the great sources of propsperity. In 1862, with copper from £110 to £93 a ton, the profit of the Knockmahon enterprises was £20,165. From 1860 to 1869, Min. Stat. record outputs of some 5,000 to 7,000 tons of ore per annum, with about 10 per cent. of copper. In 1869, a year of depression, searches were made at **Annestown** (6" 25 S.E.), 2½ miles east of Dunahrattan Bay, where lead ore had been prospected by the Company in 1837-8. Mem. 167 etc., p. 56, mentions previous searches here for copper, abandoned before 1865. A new lode was worked at Kilduane in 1872.

In 1875, the M.C.I. alleged the low price of copper (£90 a ton, however, was obtained) as a cause of loss ; but the quantity of ore available seems to have been greatly diminished. In 1876 an Inspection Committee reported to the Company that from 1826 to 1875 the losses were £43,757 and the profits £331,126. Continuous profits had been made from June 1834 to June 1846, and from June 1851 to June 1868. The mines since then proved "much less productive." Further exploration was recommended in **North Tankardstown** in 1877 ; but the Company, evidently acting on the opinion of the body of shareholders expressed in November 1876, endeavoured to sell as a going concern, while losing over £2,000 a year. Final explorations in old as well as untried ground having yielded no result, operations were closed in 1878. A plan of North Tankardstown and Glen was deposited in the Home Office series as abandoned in 1879 (No. 965). In 1879 the machinery was sold, and the Tankardstown lease was surrendered. In

1882 the Company retained only the royalty of the townland of Knockmahon.

The property was again taken up by the Bonmahon Copper Mines Development Syndicate about 1904, and the Min. Stat. give outputs from Tankardstown of 270 tons of ore, with 5 per cent. copper, in 1906, and 100 tons, with 4 per cent. copper, in 1907. Later reports, however, do not seem to have been favourable.

Mr. H. J. Daly reported on Bunmahon and Tankardstown to the Depart. Min. Resources in 1917. A general plan of the Knockmahon Mines, showing the shafts, on the scale of 24 inches to one mile, is preserved in the Home Office series of records of Abandoned Mines (R 397, dated 1845, by Warington Smyth).

Killelton (Lady's Cove). 1″ 178. 6″ Waterford 32 N.E. Copper was at one time worked here, 3 miles west of Bunmahon (Griffith Map and 1861, p. 153, and MS. 6″ Geol. Surv. map).

Woodstown. 1″ 179. 6″ Waterford 25 S.E. The MS. 6″ Geol. Surv. map shows three shafts in this townland, as " old workings for copper." The mine was known to Griffith (Map and 1861, p. 152). This is possibly the Woodstown referred to by Stewart (p. 133), though under lead ore.

Valencia. 1″ 182. 6″ Kerry 87 N.E. This mine was opened on a lode on the south side of the island, east of Clynacartan village, on Foilhomurrum Bay, There is no mention of it in Kinahan, and Memoir 182, 183 and 190 (1861) was probably written before development had occurred. A sign for copper was placed on the 1″ Geological sheet published in 1859, and the lode is marked by a gold line in the revised issue of 1879. Min. Stat. give 65 tons of ore, containing $23\frac{3}{8}$ per cent. copper, as raised in 1861, and $34\frac{3}{4}$ tons, with $24\frac{1}{8}$ per cent. copper, in 1862. There seems to be no further record.

Ardtully (Cloontoo, *Clontua*). 1″ 184. 6″ Kerry 93 N.E. Mr. W. B. Wright, of the Geol. Survey (1920), furnishes the following notes on the group of mines brought together under the name Ardtully on the map :—

" The Roughty Valley above Kenmare appears to contain in all twelve well marked lodes. Of these eight are copper lodes and four lead lodes, thus :—

Copper Lodes.	Lead Lodes.
Ardtully Lode.	The Galena Lode, Cloontoo.
The Forge Lode, Cloontoo.	Shanagarry Main Lode.
Mamby's Lode, Cloontoo.	Shanagarry South Lode.
Trinity Lode.	Killowen Lode.
Trinity South Lode.	
The three Slaheny Lodes.	

The most extensive of the old workings are on the Ardtully Lode, the Forge Lode and the Shangarry Main Lode. Trial pits or open workings appear to have been made on Mamby's Lode, the Trinity Lode, the Galena Lode at Cloontoo, Shanagarry South Lode and the Killowen Lode; but the amount of ore removed in these cases cannot have been great. It is doubtful if Trinity South Lode was ever tried, and the Slaheny Lodes do not appear to have been touched. The workings on the Forge and Ardtully Lodes have recently been re-opened by the South of Ireland Mining Company, but the operations do not appear to have been successful.

"The lead lodes lie entirely within the limestone; the copper lodes, on the other hand, occur either in the limestone, the Carboniferous Slate, or the Old Red Sandstone, and may break across from one into the other, as in the case of the Ardtully Lode."

Weaver (1838, p. 28) says that by his time "Clontua" had been tested by open-casts, none of which penetrated below 7 fathoms, and by a new shaft reaching 13 fathoms. No early adventurer had found evidence of any permanent extension of the ores. In Kane's time (1845, p. 196) Ardtully was taken up by the Kenmare Mining Association, and was raising copper pyrites "of moderate richness," about 100 persons being employed here and on the lead mines. "Gray copper ore" (tetrahedrite) is said to have been common as well as bornite. S. Haughton, in "Notes on Irish Mines," G.S.D., vol. 6, p. 211 (1855), describes the lode, and gives an analysis of the tetrahedrite. Before 1859 (Mem. 184, p. 37) the shaft reached 60 fathoms; Haughton gives 66 fathoms in 1855.

The recent working is recorded in Min. Stat., where an output of 59 tons of ore, with 4 per cent. copper, is given for 1911. Ardtully was reported on by Mr. H. J. Daly for the Department of Mineral Resources in 1917.

Carrigcrohane (Coad Mines, including **Behaghane** and **Garrough**). 1" 191. 6" Kerry 106 N.E. The quartz lodes are in Old Red Sandstone on Coad Mountain, 7 miles S.W. of Sneem. In the Home Office records of abandoned mines (R 58) there is an interesting map, entered by accident for some time under Ballycummisk. This was made on the 6" sheet by John Calvert, who writes, "these lodes, surveyed and laid down for first time by John Calvert, F.G.S., C.E., May, 1858." A copper mine is marked on each of the adjacent townlands of Garrough and Behaghane. These appear also on the engraved 6" Ordnance Survey map of 1846, on Griffith's Map, and in his Catalogue of Mines (1861, p. 147). The Geological Survey probably mapped these lodes before Calvert, since 1" 191 was published, geologically coloured, in 1858. Memoir 182, 183 and 190, p. 34 (1861), describes a quartz lode 8 feet wide, bearing E. 10° N., and traceable, with variable width, in the same direction for nearly 2 miles. Three old shafts are shown on the MS. 6" sheet, and the name "Carrigcrohane Copper Mines" was engraved upon the map by the Ordnance Survey in 1846. The earlier period of activity of the mines seems uncertain. They appear as reopened in Min. Stat. 1907; but only 1 ton of ore was raised, possibly as a trial sample. The copper content was 13 per cent.

THE BEARHAVEN MINES.

1" 198. 6" Cork 114 and 127. These mines form a sort of outlier of the Cork copper mining area. The lodes are in Old Red Sandstone grits and slates on the west slope of the Slieve Miskish Mountains, and descend to the seashore of the promontory on which Castletown Bearhaven stands. The spelling *Berehaven*, though frequently used, is less correct, the name being derived from the Iberian Princess Beara, wife of Owen More of Munster (P. W. Joyce, "Irish Names and Places," ed. 7, vol. 1, p. 134). The first records of the mines are under the name of Allihies; but the name Berehaven Mine was adopted when the workings shifted from Allihies townland (Kane 1844, p. 185). Kane attributes the discovery of the ore to Colonel Hall (see Glandore); Kinahan (p. 78) repeats this, with the date 1810; Weaver (1838, p. 27) says "discovered by a Wicklow miner in the year 1812." The workings were down to 50 fathoms before 1835. Hunt's record in Memoir Geol. Surv. Great Britain, vol. 2, pt. 2, p. 713, gives 88,636 tons of ore as raised between the opening in 1813 and 1842, the annual average being 3,300 tons, and the maximum 7,288 in 1835. Weaver states that the price in Swansea was £10 per ton.

Hunt (1848) gives, in addition to the Allihies output for 1842, 3,613 tons in that year from " Bearhaven." Subsequent records, for the reason stated above, appear under the name of Bearhaven, and show that some 5,000 tons annually were raised from 1845 to 1861 (6,969 tons in 1851), the ore averaging near 10 per cent. copper.

The following figures are selected from the records of the sales at public ticketings at Swansea in Min. Stat.:—1863, 8,358 tons of ore, value £72,118, the price of copper being near £100 per ton; 1870, 1,807 tons only, $8\frac{5}{8}$ per cent. copper; 1871, 3,876 tons, an output doubtless stimulated by the rise of copper during the year from £73 to £93.

About 1880 the output comes well under 2,000 tons per annum, sinking to 104 tons in 1883, when copper mining was ceasing to be profitable throughout Ireland.

The Berehaven Mining Comapny, 1 Foster Place, Dublin, worked the mines at least as far back as 1860, but went into liquidation in 1884. Mr. Puxley seems to have been responsible from 1863-7. The South Berehaven Mining Company of 1885, with its office at 23 Leadenhall Street, London, seems to have traded on the name of the northern locality, but concerned itself with Kilcrohane and Durrus, on the southern shore of Bantry Bay. A Company called the Berehaven Copper Mines, Ltd., of 19 St. Swithin's Lane, London, did some work at the Mountain Mine and Dooneen, Allihies, in 1899 to 1908, and also on the dumps of Crookhaven on Mizen Head. The whole Allihies district was once more developed by Messrs. J. and J. Kelly, The Allihies Copper Mines, Allihies, in 1918. The Mountain lode and the Commanmoor lode north of it were both being worked in 1920.

Jos. Dickinson, one of H.M. Inspectors of Mines, deposited in the Home Office a valuable MS. " Memorandum *in re* Records Plans of the Berehaven Mines, situated at Allihies, County Cork, Ireland, 24th October, 1885." Quotations from this will be given under the short reference, " Dickinson MS." This MS. names the shafts in considerable detail, and the sites of many can be traced on the published 6″ maps. The MS. 6″ sheets in the Geological Survey Office in Dublin show the lodes as mapped by Sir W. W. Smyth, J. O'Kelly and G. H. Kinahan. A brief but very valuable account of the mines was furnished to the Geological Survey by Sir W. W. Smyth (Mem. 197, 198, p. 30, 1860). Mr. H. J. Daly, reporting to the Controller of the Department for the development of Mineral Resources, deposited plans in the Survey Office in Dublin in 1918, and Mr. John Kelly and Mr. R. B. Ash very kindly supplied a map showing the lodes, and gave an account of recent development in 1920. Copper pyrites is the prevalent ore, and the gangue of all the lodes is quartz.

The original **Allihies** mine (the " Old Lode "; also known as **Dooneen** and *Doneen*; 6″ Cork 114 S.W.) was worked where a lode runs eastward into the townland of Allihies from its exposure on the sea-cliffs. The name Dooneen appears here, with the site of the shaft, on the 6″ Ordnance Survey map of 1901. The shaft reached some 60 fathoms before it was first abandoned in 1835. An old plan of Dooneen was deposited in the Mining Record Office (Home Office Series, R 65), and a plan and sections (892) showing work down to 90 fathoms in 1878. Dickinson MS. gives a little sketch-map of the shafts on the coast-line at Dooneen.

Operations were shifted at an early date to the **Mountain** mine in Cloan townland (6″ 114 S.W.) on a remarkable E. and W. quartz lode in places 60 feet wide, unpromising on the surface, but remarkably rich in depth. Smyth's account of its development certainly offers encouragement to prospectors elsewhere. A mining village, Allihies, sprang up here, on the slope that rises steeply from the sea to 550 feet at the Mountain mine. Kane (1844, p. 185) says that the working in his time was 760 feet long and down to 852 ft.

(142 fathoms). The lode is being actively worked by Messrs. Kelly at the present time, and the **Commanmoor** lode, which lies 200 yards north of it, is also being dealt with through an adit from the Mountain mine. This is, no doubt, the *Kamanmore* lode of Dickinson MS., which, in his time, had been excavated by a trial shaft 9 fathoms deep, about half a mile to the west of the Mountain mine. A disused mine appears here on the 6" map of 1901, on the westerly prolongation now known as the **Marion** lode (Messrs. Kelly's map).

Plans and sections of the Mountain mine are in the Home Office (R 65 and 1391). Workings on the E. and W. lode are shown as 12 to 15 feet wide, and rarely 24 feet wide; on the N. and S. lode up to 45 feet wide, the greater breadth being towards the south. Two hundred and forty fathoms were reached in both lodes. There is also a plan of the *Cluin* mine, Allihies (5186), deposited in 1908; this may be one of the mines in Cloan townland.

The **Caminches** mine (*Kamminches* of Dickinson MS.) is named from its position at the north end of the great "Caminches lode" that originates in the east of Cloan, and runs a little W. of S. for 1½ miles across the townlands of Caminches, Kealoge and Cahermeelehoe. The Caminches shaft is in Cloan (6" 114 S.W.). In Smyth's description (Mem. 197, 198, p. 31) the direction N.N.W. should probably read N.N.E. The mine was worked to 130 fathoms by Mr. Puxley about 1840 (Dickinson MS.). After a good period of productivity, the ore died out at 162 fathoms. Kane (1845, p. 194) found the Mountain and Caminches mines employing about 1,000 persons, 700 of whom were adults. He gives the working on the "Caminch vein" as 570 feet long and 912 feet in depth, the engine-shaft sinking 60 feet more. The lode was 1 to 12 feet wide. A large but somewhat damaged section of Caminches is in the Home Office (R 20).

The **Coom** mine, 6" 114 S.W., lay east of the Caminches mine on the western border of Coom townland. Its engine-shaft was known as Bewley's shaft. Compressed air drills were used here, but only 70 or 80 tons of ore were raised (Dickinson MS.). A plan in the Home Office (No. 1805; see also R 65 and 365) shows Bewley's shaft reaching 50 fathoms, and Bewley's east shaft 40 fathoms. Dickinson records a separate third shaft to the east sunk to only 8 fathoms.

Where the Caminches lode crosses Kealoge, south of Caminches, extensive and deep workings were carried to 100 fathoms in the **Kealoge** mine (*Kealogue* and *Keallogue*; 6" 114 S.W.; Home Office Plans R 65, 281, with a section down to 240 fathoms, including Reed's shaft, deposited 1875). Smyth speaks of this as "a very rich mine." **New Keallogue** (Home Office plans and sections Nos. 1805 and 1391; also Dickinson MS.) was an exploration on the west, reaching 50 fathoms.

Dickinson (MS.) records that, prior to the operations of the Berehaven Mining Company, Mr. Puxley worked Sweeney's shaft on the Caminches lode to about 100 fathoms. This shaft (6" 127 N.W.) is in the extreme south of Kealoge. The MS. 6" map of the Geological Survey marks Reed's shaft (Home Office section 281) as a little farther south, where "Copper Mines" is engraved on the map. These mines are probably the **Cahermeeleboe** mine of Griffith (1861, p. 142). Tobin's shaft, also worked by Puxley, was about 100 yards N. of Sweeney's.

The **Urhin** mine (*Irhin* of Smyth, Mem. 197, 198, p. 32; *Ihron* and *Ihrone* of Dickinson MS.) is in the south of Urhin townland (6" 114 S.E.). It lies some 550 feet up on the Knockoura slope, 1½ miles N.E. of the Mountain mine. Smyth (Mem. 197, 198, p. 33) stated that it raised great expectations; but Dickinson (MS.) says that it was abandoned about 1860, after £40,000 had been spent on it. He shows, in a rough map, that it had an engine-shaft said to be 70 fathoms deep, and four other shafts, with a dressing floor lower down the mountain side, and was approached by "a bad and steep path and bridle road" from the Mountain mine.

Tragh-na-mban (*Tragh-nam-ban*), known also (Dickinson MS.) as "Kinnahan's Shaft," was close to the shore, some two miles from Allihies village (6" Cork 127 N.W.). Dickinson says, in 1885, that it was then the most recently sunk shaft, but that only about 15 tons of ore had been sent from it. The Home Office plan (No. 1804, Oct. 1885) records small workings to a depth of 25 fathoms. Mr. Ash also informs me that very little work was done here.

The Allihies district was reported on by Mr. H. J. Daly to the Controller of the Department for the development of Mineral Resources in 1917.

The Mining District of West Carbery.

The country south of Bantry Bay, extending westward in the two long promontories that bound Dunmanus Bay, and eastwards towards Skibbereen, in fact, the old barony of West Carbery, has long been known for a wide diffusion of copper ores. The general geological structure of the district reveals two great anticlines, along which the Old Red Sandstone has been bared from its covering of Carboniferous Slate, while the slate remains as a fringe to the southern shore of Bantry Bay, in the second syncline along which the "ria" of Dunmanus Bay has been formed by subsidence, and in a third syncline coming down through the East Division of the barony and broken into two by a small anticline, where the drowned valley of Roaring Water Bay cuts across the strike. The

Yellow Sandstone Beds (Kiltorcan Series) were traced by Jukes as an "Upper Old Red Sandstone" series, between the typical and coarser Old Red Sandstone and the Carboniferous Slate, and the copper-bearing beds occur near the base of this series or a little below it. R. Griffith recognised this fact in his early explorations. In 1819, in 1821, and onwards, he visited the West Carbery district, and he observed ("On the Copper Beds of the South Coast of the County of Cork." G.S.D., vol. 6, p. 201, 1855) that the position of the copper beds "is generally very near to the outgoing of the Yellow Sandstone," the ore-bearing beds generally revealing themselves by a bright green tint. Further references to this paper will be given under the short heading "Griffith, 1855."

In commenting on Griffith's paper, as President of the Geological Society of Dublin, J. B. Jukes (*ibid.*, p. 267) corroborated, on behalf of his colleagues Willson and Wyley, this important stratigraphical observation, and he remarked that there would appear " to have been an original sedimentary deposition of copper and lead ores contemporaneously with this set of rocks, whether those metallic minerals have been left in their original state, or have been subsequently segregated into cracks and fissures, and now occur as true lodes, or as lode-like beds." The observations of the Geological Survey were made in 1853 to 1856, and Jukes restates in Memoir 200, 203-5 and 199, p. 27 (1861), his contention that " there was a great mechanical deposition of copper ore in the beds formed at the bottom of the water, so that all the grits and slates were here and there impregnated with copper ore over all the district stretching from Waterford through Cork into Kerry." It may be noted that, in his reference to Waterford, Jukes cannot have intended to include the Knockmahon district, where the ore occurs in Lower Silurian strata. Continuing, Jukes pictures some of the ore as being leached out of the beds and deposited as lodes in the fissures caused by the subsequent folding of the district. The observations of Mr. T. Hallissy in 1919 on the barytes veins of southern Ireland, now in course of publication by the Geological Survey, should be consulted in connexion with this reasoning.

Jukes no doubt rightly held that the richness of the lodes at Allihies was exceptional, and that, while " it will be obvious that a large quantity of poor ore, easily accessible, may be more productive of profit than the richest ore," yet (p. 28), " the very fact of the wide diffusion of copper ore in small quantities over so large an area is against, rather than in favour of, the probability of rich mines being found." The frequency of small examples of rich ore, without further extension as profitable masses, makes the district, as he observes, most delusive to the speculator.

S. Hyde, as the result of a paper " On deep-mining with relation to the physical structure and mineral-bearing strata of the south-west of Ireland " (Quart. Journ. Geol. Soc. London, vol. 26, p. 348, 1870, published in abstract only), raised some discussion on the age and the mode of origin of the copper deposits of West Carbery and Bearhaven. David Forbes then said that he regarded the veins as post-Carboniferous (which Jukes would certainly have admitted), and, in opposition to the views of Jukes, as not filled by segregation from the Devonian strata. Here, however, he seems to have spoken without knowledge of the features in the field.

The list of mines that have been actually worked is long, but is, as Jukes urged, somewhat deceptive. In the following notes a selection has been made from Griffith's lists (1855, p. 204, and 1861, p. 141), and from Kinahan's descriptive account for the Geological Survey (Mem. 200, 203-5, 199, p. 21, 1861), and from that by Jukes of the mines from Bantry south-westward along the southern shore of Bantry Bay (Mem. 192 and 199, 1864). For convenience, the names have been placed in alphabetical order, as will be done in the case of the numerous iron mines of north-eastern Ireland. It may be noted that Ballycummisk, Cappagh, and Horse Island were formerly known as the **Audley mines** (Weaver 1838, p. 26; Griffith 1861, p. 141), from their being on Lord Audley's property. The "Audley" mine of Hunt (1848) is probably Cappagh, since the record of the output of Audley begins where that of Cappagh closes.

The development of copper mining in this district began with the recognition and raising of ore at Ballycummisk and Horse Island in 1814, at Ballydehob in 1818, and at Cappagh in 1820. When Colonel Hall sent cupriferous peat-ashes from Glandore to Swansea (see under Glandore), vein-mining seems to have been unknown in the whole county of Cork. Allihies opened in 1813. It is remarkable, however, that the indications of copper and lead ores in the county of Cork, afterwards so freely noticed, were unknown to Stewart (1800).

Ballycummisk. 1″ 199. 6″ Cork 140 S.W. The mine is marked on the Mineral Map by the coloured spot N.W. of the Horse Island mine, and below the word Sta. of Woodlands Sta. The shafts are marked on the 1″ Ordnance Survey map, north of Rossbrin Cove. Weaver (1838, p. 26) states that *Ballycomisk* was sunk 20 fathoms in a quartz lode two feet wide, associated with barytes, and with shale mixed through it. Hunt (1848) records under *Ballycummich* 16 tons of ore, raised as far back as 1814, and 42 in 1815. The next record is in Min. Stat. (*Ballycummish*) 1857, when 17½ tons were sold by private contract. The ore raised in the next few years yielded over 10 per cent. of copper, but declined towards 7½ per cent., with a maximum output of 671 tons in 1869. Memoir 200, etc. (1861) speaks of the mine as then being one of the best developed and very satisfactorily worked. Details of the nine lodes cut by the adit level are given, including one well known as the Lady's Vein. The Lady's Vein shafts are marked on the engraved 6" map ¼ mile N. of Rossbrin Cove.

The Ballycummisk Mining Company worked the mine from 1872. In 1877 only 63 tons of ore were raised, and the plans were deposited as those of an abandoned mine in the Home Office in 1878. These plans and documents (No. 833) include a large section down to 228 fathoms, dated February, 1878, and a statement that the mine was worked by the Ballycummisk Company and latterly by Mr. Samuel Hyde. In 1857 a number of plans and sections of Ballycummisk, by J. Calvert, were deposited (R 58 and R 59), with miscellaneous MS. reports and a landscape view. The most considerable report is by S. Vivian, 6th June, 1853, who describes and highly recommends the lodes. A pencil note by J. C. (John Calvert) remarks that the descriptions do not refer to the sections, and are thus " partly incomprehensible."

As already stated, Calvert's map of the Carrigcrohane mines was at one time included in R 58 in the Home Office series. For work on the eastward extension of the Ballycummisk lodes, see Cappagh.

Ballydehob. 1" 199. 6" Cork 140 N.W. This mine lies a little to the N.W. of Ballydehob village, near the head of the drowned valley that runs up from Roaring Water Bay. Hunt (1848) no doubt refers to this mine when he gives outputs from *Ballydanab* and *Ballydahab*. Hunt quotes ore raised from 1817 to 1822 (288 tons in 1820), and again in 1826 (8 tons only). Weaver (1838, p. 26) found the shaft 20 fathoms deep. Kane (1845, 192) says that the mine employed in its period of activity 200 persons; but his statement that " many thousand tons of ore were shipped to Swansea " was probably based on conversations with old miners; it hardly tallies with Hunt's total of 606 tons. The mine was abandoned because of the poverty of the lodes in depth. In 1855, however, the Mining Company of Ireland prospected here (M.C.I. Reps.), raised 30 tons with $3\frac{1}{8}$ per cent. copper in 1856 (Min. Stat.), possibly from Kilcoe (see below), and arranged for an engine in 1857. After considerable expenditure in the hope of keeping on the workings, they were reluctantly abandoned in 1860; but the M.C.I. appear as the proprietors down to 1862. William Hobson and Company are named in the Min. Stat. Lists of Mines, 1863-5. Holdsworth (" Geology etc. of Ireland," p. 39, 1857) says that the M.C.I. sank their shaft on the lode at Kilcoe (which see). This must be regarded as a separate mine, since it lies $2\frac{1}{2}$ miles to the south-east (6" Cork 140 S.E.). Mem. 200 etc., p. 25 (1861), merely speaks of the proving of two lodes at Ballydehob, and of the Kilcoe mine as proved for about 55 fathoms by the M.C.I.

Griffith (1861, p. 141) states that the Ballydehob mine was (at some time) worked by the South Cork Mining Company. As adjacent mines he names **Boleagh** (N.E.), **Cooragurteen** (N.W.), **Kilcoe** (S.E.), and **Skeaghanore** (S.E.). All these will be found marked on his Map (1855). See also Kilcoe later.

Bawnishall (*Bawneshall, Bawnies Hall, Bannishall*). 1″ 205. 6″ Cork 151 N.W. This old mine is mentioned by Griffith (1855, p. 204, and 1861, p. 143), and in Memoir 200 etc., p. 25, where a reference is given to copper being known here in 1750. It was in the south of the townland of Bawnishall, about one mile north of Toe Head. A very small output is recorded (Hunt 1848), and for two years only, 1838 and 1840. The name appears in Min. Stat. Lists of Mines 1862-5, when a "Bannishall" Mining Company seems to have existed.

Boulysallagh. 1″ 204. 6″ Cork 147 S.E. A mine was at one time worked for both copper and lead in Boulysallagh, south-west of the village of Goleen on the south side of the Mizen Head promontory (Griffith, Map; also 1855, p. 205, and 1861, p. 142, where he calls it also *West Carbery*). Memoir 200 etc., p. 23, says that two lodes were here "anciently worked to a slight extent." It may be concluded that the mine was on these lodes, which are shown near one another on the 6″ MS. map of the Geological Survey, and not on the copper lode that is marked north of the village ("Goleen Lode"), and in the same townland. Small adjacent mines are recorded by Griffith as **Callaros** (the lode runs in from the coast in Callaros Eighter, south of Goleen, and is marked "copper and barytes" by the Survey), **Spanish Cove** (copper and lead; Mem. 200 etc., p. 23), and **Kilbarry** (6″ Cork 147 S.W.). Memoir 200 etc., p. 22, says that the lode in Kilbarry was sunk to 20 fathoms. These ventures, known collectively with Crookhaven as the Crookhaven mines, were unimportant. The sites are between the names Coney Island and Crookhaven printed in colour on the Mineral Map. *Carbery West* occurs in the Min. Stat. Lists of Mines from 1862-5 (Carbery West Co.).

Brow Head. 1″ 204. 6″ Cork 152 N.W. Brow Head is one of the most southerly points of Ireland, jutting out on the east side of Barley Cove, near Mizen Head. The mine is on the same line of strike as that of Crookhaven. Griffith (Map, 1855, and 1861, p. 142) records it as **Mallavoge** (*Brown Head*), the latter name being an obvious misprint. Kinahan (p. 28) writes *Mullavoge*. In Memoir 200 etc., p. 22 (1861), Kinahan describes the lode as "formerly worked," with good pyritic ore at a depth of 30 fathoms, the inclined bed being proved down its dip for a distance of 60 fathoms. Min. Stat. give their first record in 1859 (179 tons of ore, $10\frac{1}{16}$ per cent. copper). Only 59 tons were raised in 1860; the Brow Head Company was responsible from at least 1862-5, but no further record appears until 1906 (2 tons, 22 per cent. copper). This was got by the West British Development Syndicate, 7 Tower Buildings North, Liverpool.

Mr. H. J. Daly reported on the mine to the Controller of Mineral Resources in 1917.

Cappagh (Cappaghglass). 1" 199. 6" Cork 140 S.W. This is the best known of what were called the Audley Mines. It lies E. of Ballycummisk, and is marked on the engraved 1" map nearly two miles south of Ballydehob. Griffith (1855, pp. 201 and 202) visited these mines in 1819 and 1821, and the reports of the M.C.I. (1824) show that at one time he conducted the operations at Cappagh. Hunt (1848) gives, for 1820, 30 tons; 1821, 57 tons; 1822, 33 tons. The M.C.I. took over the mine in 1824, sank to 84 fathoms between 1825 and 1830 (Griffith 1855, p. 202), and raised 448 tons in 1827, but found the lode in time unprofitable, and in 1830 surrendered the lease to Lord Audley, with the machinery, for a consideration of £12,000. The mine seems at that time to have been worked under the name of **Audley** (Hunt 1848); the records are 367 tons in 1830, 152 in 1831, and only 11 tons in 1832. Kane (1844, p. 183) describes the lode, which became tested to 120 fathoms, with levels extending over 200 fathoms. Expectations of ore at the junctions of veins having failed, working was again abandoned. In consequence, Memoir 200 etc., p. 24 (1861) dismisses the mine with a few words; but in 1863 the Cappagh Mining Company took it up (Min. Stat.), raising small outputs until 1866. From 1867 to 1872 there is no record; 35 tons of ore were raised in 1873, and work then ceased. *Great Cappagh* is given independently as a mine-name in Min. Stat. Lists of Mines 1862-5 (Cave and Company).

In the Home Office series (R 58) there is a large section of Cappagh, showing 11 levels, down to 114 fathoms, with a note that operations closed in 1873. J. Calvert also gives a large plan (date 1857), which he could not guarantee, since local information was of an interested nature (as is commonly the case in mining enterprises), and the mine had been full of water for twenty-five years. In R 58, which is a roll miscellaneous records, there are also various sections, sketches, and plans of Cappagh, two lithographs for issue to the public or to shareholders, and an undated report by P. T. Foley, Junr., to which "J. C." adds a note of warning disapproval. There are also reports by S. Vivian (1853) and M. Edwards (1854) on the **Bog Mine**, a little north of Cappagh, where lodes had been tested to 12 fathoms.

Carravilleen. 1" 199. 6" Cork 129 S.E. A small mine east of Killeen, closed before 1864. See under Kilcrohane, and Mem. 192 and 199, p. 47 (1864). Much dolomite was present.

Carrigacat. (See Dhurode.)

Coney Island (*Coony*). 1" 204. 6" Cork 148 N.E. This

small island, north of Long Island, includes a lode mentioned in Mem. 200 etc., p. 23. "Coony and Long Island" are associated in Min. Stat. Lists of Mines 1862-5, but there are no statements of output from either. Min. Stat. 1906 give 12 tons of ore with 4 per cent. copper as raised in 1906, and 15 tons of a richer quality in 1907. The record then ceases. The recent workers were the Schull Copper Mining Company, Skull, and 16 Great Winchester Street, London, E.C.

Coosheen (*Cosheen, Coshen, Cooshen*; also known as **Skull Bay**). 1″ 199. 6″ 139 S.E. and 140 S.W. Mem. 200 etc., p. 23, names this mine as *Skull Bay Mining Company's*, or Coosheen mine. It is well to note this at the outset, since official reports commonly use the name Cosheen and Coosheen from 1840 onwards, but also Skull Bay or Schull Bay from 1860 to 1865. The mine is on the east side of Skull Harbour, the spot indicating it on the Mineral Map being that vertically above the "C" of the name Coosheen. Kane (1845, p. 194) describes the mine as begun by Messrs. Connell and McMullen, of Cork, in 1839, and employing 130 persons. Mem. 200 etc., p. 23, names eleven lodes, resembling beds, as is usual in West Carbery. Most of them, in 1860, had been proved only near the surface. Hunt (1848) gives an output of 1,957 tons of ore between 1840 and 1847. Min. Stat. 1854 carry on the record, and the ore at this time gave as much as 18 per cent. of copper. The record under Skull Bay closes in 1862 with ore of a low grade. The mine is named in the Lists until 1865, and was revived from 1870-8, being, from 1875 at least, under Mr. Samuel Hyde. Its plan was deposited at the Home Office among those of abandoned mines in 1878. An earlier plan (R 58) is also preserved under the name *Schull Bay*. The mine reappears, however, in 1888 under the management of Messrs. Tribe, Clarke, Painter and Company; in 1890 it was worked in connexion with Killeen North (see Kilcrohane). It was once more worked, with a trifling output, in 1906 and 1907.

Crookhaven. 1″ 204. 6″ Cork 147 S.W. and S.E. and 152 N.W. The mine is on the narrow promontory that terminates in Streek Head, south of Crookhaven and east of Crookhaven village. Mem. 200 etc., p. 22, names and describes four lodes, on the same strike as the bed worked at the Brow Head mine, which lies $2\frac{1}{2}$ miles to the south-west. The large "Champion" quartz lode was worked by an engine-shaft in 1860 to a depth of 40 fathoms. Min. Stat. record 43 tons of ore in 1854, with $5\frac{7}{8}$ per cent. of copper; the name appears in the Lists of Mines from 1860-5 (A. C. Langton and Company), but no output is given. In 1903-5, the Berehaven Copper Mine, Ltd., of Allihies and London, worked the dumps here and near Allihies. Plans of Crookhaven were deposited in the Home Office abandoned mines series (R 397), in 1863.

The term **Crookhaven mines** was used to include a number of ventures, some, like **Kilbarry** and **Spanish Cove**, of no moment (see under Boulysallagh above; also Griffith's list, 1861, p. 142, and Kin., p. 28). Kilbarry was worked by the Kilbarry Mining Company from 1862-4.

Derreennalomane. 1" 199. 6" Cork 130 S.E. A copper lode lies on the south slope of Mount Corin, and copper ore is associated with the barytes for which the Derreennalomane workings are at present known. The locality is ½ mile north-west of the village of that name (Portiroe of the old engraved 1" map). Griffith (1861, p. 141) gives this as one of the Ballydehob mines; but no separate record of output has been traced. Memoir 200 etc., p. 21 (1861) speaks of trials only, on the lode near that containing barytes. The two lodes are marked separately on the 1" geological sheet, and Min. Stat. Lists of Mines, 1862-78, record the mine as working both copper and barytes. See under Barytes.

Derreengreanagh (Scart). 1" 199. 6" Cork 118 N.E. This lode, in Old Red Sandstone S.E. of Bantry, is marked on Griffith's Map, but is given by him as unworked (1855, p. 205, and 1861, p. 142). Griffith recorded barytes, and it has been since worked for this mineral. See under Barytes.

Derreennatra. 1" 199. 6" Cork 140 S.W. Mem. 200 etc., p. 24, speaks of a trial made here, on lodes or beds between those of Cappagh and Ballycummisk. Hunt (1848) records the actual raising of 61 ton of ore in 1843; but no progress was made.

Derrycarhoon. 1" 199. 6" Cork 131 S.E. Some work was done on this lode in the townland of Derreennaclogh near Ballydehob in 1852. This townland lies east of that of Derrycarhoon, the name of which appears in Min. Stat. Lists of Mines 1862-73 (Swanton and Company). Kinahan (pp. 6 and 121) gives an interesting account of ancient excavations here, the entrance being "smothered up by a growth of peat, over fourteen feet deep." Mr. H. J. Daly reported on the locality to the Department for the development of Mineral Resources in 1917.

Dhurode (Carrigacat). 1" 199. 6" Cork 147 N.W. This mine is marked on the Geological Survey 1" map and on the current topographical map, on the southern shore of Dunmanus Bay, at the foot of a steep descent from the Old Red Sandstone upland. It is in the joint townland of Carrigacat and Milleen, and Griffith's Map shows two mines, Carrigacat and Dhurode. There seems, however, to be only one lode. The lode of **Lackavaun** lies to the north-east, and that of **Balteen** to the south-west, both of which are given by Griffith (1861, pp. 142 and 143) as worked by mines. Hunt (1848) gives an output from Dhurode of 229 tons of ore from 1844-6. In 1850 the M.C.I.

took up the mine, but abandoned it later in the year. Something may have been done between 1860 and 1865 (Min. Stat. List of Mines). Mem. 200 etc., p. 21 (1861), speaks of small and abandoned shafts on a lode bearing E. and W. The gossan was auriferous (Griffith, 1861, p. 142). The mine was temporarily revived in 1900; the Home Office List of Mines records it as in the hands of the Dhurode Copper Company, Goleen, County Cork, in 1903 and 1906.

Glanalin (*Glenanlin*). 1″ 199. 6″ Cork 129 S.E. This mine is shown on Griffith's Map on the Sheep's Head promontory, between Killeen and Carravilleen (see under Kilcrohane). It is mentioned in Min. Stat. Lists of Mines as worked by a Glenanlin (probably Glanalin) Company from 1862-5. No output is recorded.

Gortavallig (*Gurtyvallig*). 1″ 199. 6″ Cork 138 N.W. This lode of copper pyrites and tetrahedrite runs with the characteristic easterly trend from the coast into the townland of Gortavallig, about four miles from the west end of the Sheep's Head promontory. It is marked under the hill of Cashmahignafane on the Geological Survey map. Mem. 192 and 199, p. 47 (1864) describes it as then worked by the Carberry Mining Company and as having been "worked extensively some years previous to 1854." Griffith (1855, p. 205) gives it as a worked mine, and the Carberry Mining Company held it from 1863 to 1879; but I have been unable to trace any record of output. In the Mineral Map its site is immediately under the "N" of the name Kilcrohane printed in colour. A plan (R 315, *Gortavillig*) is in the Home Office.

Hollyhill. 1″ 199. 6″ Cork 118 S.W. This mine lay south of Bantry, near and north-east of the main road to Skibbereen, in the small townland of Hollyhill; but it does not appear on the Geological Survey MS. 6″ or published 1″ maps, which show only the adjacent lead lode of Gortnacloona. Mem. 192 and 199 makes no mention either of this lode or of the Hollyhill mine. Griffith, however, placed Hollyhill on his map, and (1855, p. 205) records it as a worked mine. I have been unable to trace any output. In the map accompanying the present Memoir its site is practically hidden by the "D" of Derreengreanagh printed in red.

Horse Island. 1″ 199. 6″ Cork 149 N.W. This island is in the mouth of Roaring Water Bay, south of Ballydehob, and the mine is marked on the geological and on the current 1″ topographical map. Mem. 200 etc., p. 25, says that the lode was large, had a good appearance, but split up into strings 8 fathoms down. Its first records seem to be small outputs sold at Swansea in 1814 and 1820 (Hunt 1848).

John Forrest, in a MS. in the Home Office Records of abandoned mines (R 58), dated 20 March, 1854, mentions "old men's" workings, and a shaft of 9 fathoms, meeting "grey ore" 3 to 5 feet wide, from which much had been removed. P. J. Foley, of Cappagh (*ibid.*), writes that the main lode averages 10 feet wide, and names various shafts. In the Home Office roll R 305, J. Calvert gives a coloured map of the island and a drawing of the north coast, with plans of old works "as reported," and of works at the east end in 1856 and 1857. The ore, he says, yielded 6–12 per cent. copper, but became too difficult to work in such a shaly country. He gives a good section of the inclined lodes at the eastern end, from his own survey, noting their abandonment in November 1857. He was inclined, however, to think that the western lodes offered more promise. On a traced map, showing "the probable directions of the lodes" from his own observations, Calvert shows the Main Lode as running from end to end of the island, and describes it as "a very promising strong gossany lode." Kane (1844, p. 183) writes of a shaft from which 230 tons, yielding in some cases 55 per cent. of copper, were sold at Swansea for £2,800. On deepening this mine to 40 fathoms, the ore was found to die out, and the workers surrendered their lease. The West Cork Mining Company, financed in London, gave a high and fancy price for the mines, and raised only 173 tons of ore, ceasing work before 1844. Min. Stat. record 15 tons as sold in 1857 by private contract. This is the year of Calvert's reports, above cited, and a long lapse then ensues, though the mine is named in Min. Stat. Lists of Mines from 1860 to 1865. In 1889, 1891, 1900 and 1901 the records of output at Horse Island are associated with those of Killeen North (Kilcrohane), 13 miles north-west on the Sheep's Head promontory. The name of the mine is retained in the List of Mines for 1903.

For Horse Island mine in Co. Mayo, see Geevraun. There is also another Horse Island at the mouth of Castle Haven (6″ Cork 151 N.W.), opposite to which on the mainland the West Cork Mining Company had a store.

Kilbarry. See Boulysallagh and Crookhaven.

Kilcoe. 1″ 199. 6″ Cork 140 S.E. This mine, no trace of which, except perhaps "quartz lodes," appears on the MS. map of the Geological Survey, is marked on Griffith's Map (1855) on the east side of an inlet of Roaring Water Bay, 2½ miles S.E. of Ballydehob, lying in Kilcoe townland. In the Mineral Map its site is just south of the "B" of "Ba" printed in colour. Kilcoe is omitted from Griffith's 1855 list, but appears as a worked mine in that of 1861 (p. 141). Holdsworth ("Geology etc. of Ireland," p. 39, 1857) says that the M.C.I. shaft (of 1855 ? see Ballydehob) was sunk on Kilcoe (*Kilkoe*) lode, on the advice of Mr. Hoskin. He adds that a copper vein 7 to 10 inches thick was found here, after passing through 35 fathoms of quartz. The experience of the Mountain mine at Allihies seems thus to have been repeated to some extent. Mem. 200 etc., p. 25, says the vein was proved for about 55 fathoms by the M.C.I., who are named as holding the mine from 1862-5 (Min. Stat. Lists of Mines).

Kilcrohane (Killeen North, Killeen). 1" 199. 6" Cork 129 S.E. Several mines were worked from time to time on the northern side of the long and narrow anticline that divides Bantry from Dunmanus Bay, and it is somewhat remarkable, if the beds were impregnated at the epoch of their deposition, that the southern side, on Dunmanus Bay, should be devoid of mineral indications. Griffith quotes no mine under the name of Kilcrohane, but on his map shows in the following order, from S.W. to N.E., **Gortavallig, Killeen, Glanalin, Carravilleen** (see these names above). The three last lie in adjacent townlands along the coast, and all are copper mines (see also Griffith 1855, p. 205). Kilcrohane parish extends from the village on the south across the promontory, and there is little doubt that the mine known in literature as Kilcrohane is identical with Killeen. It stood on the coast, in the townland of Killeen North, where a lode is marked as "worked" on the MS. 6" map of the Geological Survey.

Memoir 192 and 199, p. 47 (1864), which includes the mines from Bantry to Gortavallig, states that the ore extracted at Killeen was malachite. Kinahan (p. 27) gives a Kilcrohane mine as distinct from the North and South Killeen mines; the South mine does not seem to be mentioned elsewhere. He says that there is a thick bed of iron pyrites at Kilcrohane. The copper ore was auriferous, and Min. Stat. 1885 quote 2 dwts. of gold per ton of dressed ore.

The M.C.I. worked Kilcrohane in 1837 (putting it in Kerry in their Report). They abandoned it in the latter half of 1839. After a long interval, the name Killeen appears in Min. Stat. Lists of Mines, 1862-5 (Killeen Mining Company). The mine (Min. Stat. 1884) was reopened by the South Berehaven Mining Company of London. Forty-seven tons of ore were raised in 1884, and 127 tons, with 9·5 per cent. copper and a little gold, as above mentioned, in 1885. At that date Kilcrohane was the only mine raising copper in the whole of Ireland. In Min. Stat. 1889 and 1891 a joint output is given from Killeen North and Horse Island, and in 1890 it is associated with Coosheen.

It seems desirable to retain the name Kilcrohane, known since 1837, for this mine, to avoid confusion with Killeen in Tipperary.

Long Island. 1" 204. 6" Cork 148 N.E. This mine is marked on the 1" Geological and the modern Ordnance Survey sheets, at the north-eastern extremity of Long Island, south of Skull, where the "C" of Coosheen occurs on the Mineral Map. Griffith (1855, p. 208) records it as worked before 1855. Mem. 200 etc., p. 23, merely says that trials were made here.

Min. Stat. Lists of Mines 1862-5 associate it with Coney Island (Coney and Long Island Company).

Mizen Head (Cloghane). 1″ 203. 6″ Cork 146 S.W. This mine was on a lode in Cloghane, in the inlet just south of Mizen Head, under the "C" of the symbol "Cu" on the Mineral Map. Its name appears in the Lists of Mines, 1862-5, with Swanton and Company as proprietors.

Mount Gabriel. 1″ 199. 6″ Cork 139 N.E. A copper lode seems to have been opened up here by the Mount Gabriel Mining Company from 1862 to 1874, but no output is recorded (Min. Stat. Lists of Mines). Mem. 200 etc., p. 23 (1861), mentions copper ore as here associated with the barytes. From 1894 onwards the barytes was developed (see under that mineral).

Roaring Water. 1″ 199. 6″ Cork 140 S.E. Various trials were started near and south of the former village of Roaring Water, which lay at the head of the north-eastern arm of Roaring Water Bay, south-east of Ballydehob. Kane (1845, p. 193) speaks of the Roaring Water lode as promising, and the venture as a "new mine." He was not aware that ore had then been sent to market; but Hunt (1848) records 14 tons of ore as sold in Swansea in 1844. Kane's lode was probably the most northern of the three recorded as Roaring Water lodes, Leighcloon, Laheratanvally and Kilkilleen, which are shown in the order of these townlands from north to south on Griffith's Map (1855). See also Griffith 1855, p. 204, where he gives these lodes as unworked; he repeats this statement by not writing the names in italics in his paper of 1861 (p. 142), though he there calls them mines. Kinahan (p. 27) also speaks of these three lodes as the Roaring Water mines, and gives lead as associated with the copper. The **Leighcloon** lode is the only one recorded on the MS. 6″ map of the Geological Survey. Its site is just above the "C" of *Church* printed at the N.E. inlet of the bay on the Mineral Map. The plan of "Roaring Water Mine" deposited in the Home Office (R 10, 1863) is of Leighcloon; the ore found a convenient quay at the mouth of the Roaring Water River. Leighcloon mine was thus in existence from about 1844 to 1863. "Roaring Water" appears in the Lists of Mines from 1862 to 1874.

Skull. 1″ 199. 6″ 148 N.E. Griffith's Map shows the mine of this name about a mile south-west of Skull, on the west side of Skull Harbour and north of the outcrop of the Kiltorcan series. The "Skull Bay mine," as we have seen, is the Coosheen mine on the opposite side of the harbour. No trace, however, appears of the Skull mine on the published or MS. maps of the Geological Survey, except a note " trial for copper" north of Carrigrowhig at the south point of the harbour. Mem. 200 etc. has no reference, though the mine is given in Hunt (1848) as producing 35 tons of ore in 1841 and 84 tons in 1843. Kane (1844, p. 184) writes of the

Skull mine as a going concern, but says that it was on "an island of that name." No such island can be traced. The true site is, no doubt, that given on the Ordnance Survey 1 : 2500 (25″) map, Cork 148 4, which shows a "disused shaft" on the spot indicated by Griffith.

Griffith (1855, p. 205) says that there was also a mine at **Castlepoint** (6″ Cork 148 S.W.), which is indicated on his Map, but is no longer traceable.

Skull Bay. See Coosheen.

Glandore Mines. 1″ 200. 6″ Cork 142 N.E. and 143 N.W. These mines, on the long lode running eastward from the head of Glandore Harbour almost to Ross Carbery, are of primary interest as a source of Manganese ore, and they will be dealt with under that heading and also under iron. But copper ore has been raised from most of them, the names of individual mines being given by Griffith, 1861, p. 143, and on his Map of 1855. It is possible that the output from "*Landore*" in Hunt 1848, may be from the Glandore district. There is, however, only one entry, 25 tons of ore in 1825.

It is noteworthy that the first recorded attempt at utilising the copper ores of south-west Ireland was made at Glandore Harbour about 1813. H. Townsend (Stat. Surv. Cork, p. 53, and Addenda, p. 156, 1815) records the discovery of copper in the ashes of peat-bogs here and near Castle Freke. Colonel Hall had the ashes collected and sent to Swansea, where they fetched £10 to £12 a ton. Townsend connects the occurrence at Glandore with ore-veins in the adjacent hill, but adds that the country affords "few indications of metallic possession." It seems to have required the eye of a man like Griffith to realise the possibilities of the district, and Townsend remained on the edge of considerable discoveries.

The mysterious entry, **Burnt Irish,** in Hunt's records (1848) is probably explained by the peat-ashes of Glandore. If this is so, the record as "copper ore" sold at Swansea is not quite comparable with that from the bogs, since some sulphur must have been driven off in burning. The figures at Swansea are 1814, 27 tons; 1815, 24 tons; 1818, 95 tons; and 1819, 167 tons. The bog may then have been worked out.

Duneen Bay (Duneen, Muckruss Head). 1″ 201. 6″ Cork 144 N.E. This mine, on a promontory at the west side of Clonakilty Bay, has become associated with the barytes industry. See Barytes. But Weaver, 1838, p. 25, writes of old works here, close to the sea in *Doneen* Inlet, which had as their apparent object "a bed in the slate, between one and two feet wide, composed of sparry iron ore and quartz, with disseminated galena and pyrites of copper and iron." Griffith's Map, 1855, marks Duneen as a lead and copper mine.

I have been unable to trace the following copper mines mentioned in the statistics officially published:—

Cloncurry, which was working in 1872-6, with small outputs; and these given by Hunt (1848), *Malagow, Nanmor, Conham, Knockmalaur, Fingal, Shionagree. Irish Consols* is named in Min. Stat. Lists of Mines from 1862 to 1865 as worked by the Irish Consols Company.

CHAPTER VI.

FELSPAR (FELDSPAR).

The localities where "Felspar" has been inserted on the Mineral Map are those in the neighbourhood of Belleek; but a number of other occurrences that may prove to be of commercial value have been recently visited and recorded by the Geological Survey. The veins of coarse granite that are now generally, if incorrectly, known as pegmatites, consist mainly of potassium-felspar and quartz, with only a small proportion of mica and occasionally with some tourmaline. Their coarseness of grain may allow of the separation of the mineral constituents by picking, after the rock has been quarried and broken up; in other cases, true pegmatitic intergrowth of the quartz and felspar occurs, rendering separation impracticable. The intergrowth increases the percentage of silica, but reduces the potash-content to a noteworthy degree. The veins must naturally be several feet in width to render quarrying profitable. Their materials are of value not only for the porcelain industries, but in view of the possible extraction from them of potash for agricultural use.

Dooey Point. 1″ 15. 6″ Donegal 57 S.W. and 65 N.W. Mr. E. St. J. Lyburn refers to the numerous pegmatite dykes in the district, on the east side of Gweebarra Bay, as worthy of consideration (Journ. Depart. Agric. and T. I., vol. 16, p. 627, 1916). There is easy access to the sea.

Cleengort and **Derryloaghan.** 1″ 15. 6″ Donegal 65 N.E. These townlands are on the south side of the Gweebarra, east of the iron bridge and north of Glenties, and their pegmatite dykes seem coarse enough to deserve attention. They cut the gneissic and schistose complex of the district.

Belleek District. 1″ 32.

The well-known potteries of Belleek in the county of Fermanagh were originally supplied from veins that cut the gneiss of the Precambrian district to the north; but the quantity of felspar required was never very large, and it became possible to import selected material from Cornwall and Scandinavia. The veins at **Larkhill**, and southward to Castle Caldwell railway station (1″ 32. 6″ Fermanagh 3 S.E. and 8 N.E.), offer, however, considerable possibilities, and felspar was being raised in this area in 1920.

The veins near **Lough Unshin** (1″ 32. 6″ Donegal 108 N.W.) are of similar character and have also been prospected.

A report on the pegmatites of the Belleek district, prepared by Mr. W. B. Wright on behalf of the Geological Survey of Ireland, has been printed in Prof. P. G. H. Boswell's "Supplementary Memoir on British Resources of Sands and Rocks used in Glass-manufacture" (Ministry of Munitions of War and Imperial College of Science, published by Longmans, Green and Company, 1917). Prof. Boswell gives three analyses of the Belleek rock, by Messrs. Harwood and Eldridge, showing respectively 10·48, 10·14 and 13·08 per cent. of potash. S. Haughton ("On graphic feldspar, from County Donegal," R.G.S.I., vol. 4, p. 225, 1877) analysed the "graphic orthoclase" used in Belleek pottery in 1876, and calculated the composition as quartz 4·7 and orthoclase 95·3 per cent. The potash was 12·25 per cent.

Termoncarragh. 1″ 51. 6″ Mayo. 9 N.W. The gneissic granite of the Belmullet district is traversed by a large number of bands and irregular masses of pegmatite. Some of the best occurrences are at the end of the peninsula north of Portnafrancagh, where the coast is, moreover, moderately low. At Scotch Port, ¾ mile distant, small vessels could be loaded. This locality seems distinctly favourable, should the demand for felspathic rock continue.

Doolough Peninsula. 1″ 51. 6″ Mayo 25 N.W. The peninsula terminating on the north-west in Doolough Point, on the east shore of Blacksod Bay, and some five miles south of Belmullet, contains a number of dykes penetrating the gneiss, which offer a hopeful field for exploration. The lowlying nature of the ground, however, may render quarrying difficult.

CHAPTER VII.

GOLD.

For a general review of the gold-resources of Ireland, see J. Calvert, "The Gold Rocks of Great Britain and Ireland" (1853).

Gold Mines Valley. 1" 139. 6" Wicklow 39 S.E. and 40 S.W. and N.W. The alluvial deposits in the neighbourhood of Woodenbridge, Co. Wicklow, probably supplied the gold of Ireland in prehistoric times. The magnificent collection in the National Museum, Dublin, provides sufficient evidence of the former importance of the industry, which, no doubt, was one of the first causes that brought merchants from Gaul and Iberia to the western isle. Ireland was still looked on as a gold-providing country by the unknown economist who wrote "The Libelle of Englyshe Polycye" in 1436. The Irish, he remarks, had no skill in mining; but

". . . if we had their peace and good will
To myne and fine, and metal for to pure,
In wilde Irish might we finde the cure,
As in London saith a Juellere,
Which brought from thence golde oore to us here,
Whereof was fyned mettal goode and clene,
As they touch, no better could be seene."*

J. Knight Boswell (R.G.S.I., vol. i, p. 99, 1865) has recorded how he heard from the actual finders that a nugget weighing 1½ lbs., discovered in the Gold Mines Valley, was sold to a Dublin jeweller about 1790, and that this called attention to the district.

The revival of alluvial working in the Ballinvally stream (or Gold Mines River) below Croghan Kinshelagh in 1795 and 1796, and the attempts to find gold *in situ* by excavations in the quartz veins of Croghan Kinshelagh itself, are well described in a Report by A. Mills, T. King, and J. (probably T.) Weaver, made to the Lord Lieutenant by command, and signed at Cronebane, August 21st, 1801. The Dublin Society (now Royal Dublin Society) obtained permission to print this report in its Transactions, vol. 2, part 2, p. 131 (1802), and R. Kirwan (p. 150) appended some general remarks. Abraham Mills had already described the occurrence of gold above Woodenbridge in a paper entitled "A mineralogical account of the native gold lately discovered in Ireland," read before the Royal Society of London on 17th December, 1795, and printed in the Phil. Trans. for 1796, p. 38; this was

* In R. Hakluyt, "The principal Navigations etc. of the English Nation," Everyman edition, vol. i, p. 191.

reprinted, with its accompanying map, by the Dublin Society (Trans. D.S., vol. 2, part 1, p. 454, 1801). Mills records that a specimen of 24 grains of gold was assayed by Weaver, and gave 22·16 grains of gold and 1·23 grains of silver.

R. Fraser (Stat. Surv., " General view of the agriculture and mineralogy of the County Wicklow," p. 19, 1801) describes the gold deposit generally.

Mills made a second Report, including notes on other mineral occurrences in the district, on 8th March, 1802 (Trans. Dublin Soc., vol. 3, p. 81, 1803), and T. Weaver (1819, pp. 208-213), who had charge of the systematic trenching and tunnelling operations, usefully reviewed the whole matter at a later date. He gives a coloured map (plate 11), the peculiar orientation of which must be noted by the user.

William Mallet described the " Minerals of the auriferous districts of Wicklow " in 1851 (G.S.D., vol. 4, p. 269), and showed the association of cassiterite, wolfram, and perhaps platinum, in the alluvium. Most of the minerals cited were known to Weaver in this area (1819, p. 209). Smyth summarised Weaver's account in his valuable memoir on the " Mines of Wicklow and Wexford," p. 400 (1853). Gerrard A. Kinahan contributed an important paper in 1882 " On the occurrence and winning of gold in Ireland " (R.G.S.I., vol. 6, p. 135, 1886), and G. H. Kinahan called attention a year later to the large extent of unworked alluvium (*ibid.*, p. 207). E. St. John Lyburn ("Prospecting for gold in County Wicklow, and an examination of Irish rocks for gold and silver," Proc. Royal Dublin Society, vol. 9, p. 422, 1901) assayed a large number of samples collected by himself in this district, and included the Ovoca belt in his researches. The results were mainly negative ; but a quartz vein on the summit of Croghan Kinshelagh yielded 4 dwts. of gold per ton (p. 424).

Several good nuggets were discovered in the Gold Mines Valley, and are now in various museums. Specimens were presented to the Dublin Society, and these and several models were brought together by the care of Dr. V. Ball in the Irish Mineral Collection of the National Museum of Science and Art, Dublin. Ball, in a carefully reasoned paper (" On the gold nuggets found in the County Wicklow," Proc. R. Dub. Soc., vol. 8, p. 311, 1895) traced the history of a large nugget, weighing 22 oz. Troy, which was probably found in 1795. He concludes (p. 318) that it was presented by Abraham Coates, Port Surveyor of Wicklow, to King George III. The model in the National Museum was made from one in the possession of Dr. W. Frazer (not Fraser, as in Ball's paper), a well-known Dublin antiquary.

John Calvert (*op. cit.*, p. 175, 1853) mentions a nugget of 40 oz. as exhibited in 1844. V. Ball, however, (*op. cit.*, p. 320) shows that the figure should have been 4 oz., the mistake having occurred in a newspaper report. J. Holdsworth (" Geol. etc,

GYPSUM 65

of Ireland," p. 27, 1857) mentions two nuggets as found "in the course of the past year." The larger "is said to have weighed two pounds troy." Gerrard Kinahan (*op. cit.*, p. 147, 1882) reports a nugget of 24 oz. as found by peasants about 1857. It seems strange that no further record of so remarkable an occurrence should have been traced.

Ballinvally appears as the name of a gold mine in Min. Stat. Lists of Mines 1877-9, and in Min. Stat. 1882; but no output is recorded. F. Acheson is given as responsible. This enterprise also appears (1878-9) as *Wicklow Gold*.

CHAPTER VIII.

GYPSUM.

The Triassic strata of Ireland, which were formed, like those of England, under conditions of aridity varied by seasonal rains, contain beds both of gypsum and rock-salt, deposited in desiccating lakes. These strata, however, have suffered from prolonged denudation and, like those of Jurassic and Cretaceous age, now remain only in the north-east of the country. Gypsum has been worked at a few points.

Belfast. 1" 36. 6" Antrim 60 N.E. and S.E. The brickyards on the west side of the city have yielded a fair amount of gypsum, which is in irregular and sometimes crumpled bands in the Triassic clays. It is gathered by picking from the spoil-banks (Mem. Geol. of country around Belfast, p. 123, 1904). The industry was in existence in 1812 (Dubourdieu, Stat. Surv. Antrim, p. 73, 1812).

CARRICKMACROSS AND KINGSCOURT AREA.

1" 70. 6" Monaghan 31 S.W.; and 1" 81. 6" Meath 2. S.E.

As shown on the Mineral Map, gypsum has been excavated at the northern and southern ends of the Triassic outlier that extends southwards from Carrickmacross. The deposit is in places 60 feet thick, and was know to Griffith (1861, p. 150) through work done by C. T. Shirley in **Knocknacran East** (6" Monaghan 31 S.W.). The following is an extract from a Report made for the Geological Survey by Mr. T. Hallissy in 1920:—

"The principal deposit of the mineral is found in the counties of Monaghan and Meath; it occurs in the Keuper

Marls, which extend in a narrow strip, for about six miles, from about three miles S.W. of Carrickmacross to about three miles S. of Kingscourt. In the northern part of the area the bed has been proved to be about 60 feet thick, and where it outcrops at Lisnabo, S. of Kingscourt, more than 5 feet in thickness of the bed can be seen. Between Kilmainham and Kingscourt, a pit and three borings were made for gypsum in 1879, and different beds of the mineral were encountered near the surface, varying in thickness from $3\frac{1}{2}$ to 17 feet." Some of the gypsum is stained with iron hydroxide.

A plan of a mine at **Drumgill**, near Kingscourt, is in the Home Office series as abandoned in 1886 (No. 1829). A note on it by J. Dickinson says, "This mine was worked for marl used in making red bricks, and some gypsum was got where the 21 feet of gypsum was sunk into by an underground pit. The gypsum was reddish coloured. Scarcely any of the pillars was worked." Another plan (No. 2504), deposited in 1890, shows deeper workings at Drumgill, the pit being "90 yards deep, all in red clay overlying gypsum." This pit was worked by Thompson Brothers, and closed on July 12th, 1890.

New trial workings have been recently made in this area by the Farney Development Company, and gypsum is now being raised. See also Barytes.

Coagh. 1" 27. 6" Tyrone 30 S.E. E. T. Hardman (R.G.S.I., vol. 3, p. 87, 1873) drew attention in 1872 to an occurrence of gypsum in the Keuper Marls about $\frac{3}{4}$ mile S. of Coagh, on the left bank of the Ballinderry River. Trials for coal, naturally fruitless, were made here in the Triassic shales; but gypsum was found, apparently of a fair white quality. It seems, however, to have been in comparatively small lumps, and the locality remains practically untested.

CHAPTER IX.

IRON.

The introduction of iron weapons into Ireland is believed to have been due to the "Celtic" invaders in about the sixth century B.C., and from that time onwards local ores were no doubt recognised and smelted. So long as the forests could supply sufficient fuel, iron ores were freely mined, and the traces of old "bloomeries" have given rise to traditions of exceptional supplies in certain districts. Gerard Boate ("Ireland's Naturall History," pp. 125-140, 1652) paid special attention to the sources of iron and to iron-works erected in Ireland by the English, and his description contains much of interest in connexion with these bloomeries, in which smelting by charcoal was carried on. Kinahan (p. 109) gives an interesting historical account of iron-mining in the county of Wicklow, which has applications to the country as a whole.

At the present time, the lateritic ores of the north-eastern area are alone exploited for smelting ; but in 1915, when a knowledge of all possible sources of iron in the British Isles became especially urgent, an account of the more important occurrences in Ireland was supplied by the Geological Survey for the information of the Advisory Council of the Department of Scientific and Industrial Research. This summary was published, together with notes obtained from other sources, in a "Report on the Sources and Production of Iron and other Metalliferous Ores used in the Iron and Steel Industry" (H.M. Stationery Office, "Rep. on the Resources, etc.," 1917, and 2nd edition, "Rep. on the Sources, etc.," 1918). Any differences between the present notes and the previous account may be taken as corrections of the information then brought together, since the Geological Survey had not the opportunity of revising their contribution to the published Report.

The term "limonitic ore" has been used in the present Memoir, rather than "limonite" or "brown haematite," owing to the evidence brought forward by Posnjak and Merwin (Amer. Journ. Sci., vol. 47, p. 311, 1919) to the effect that no definite species exists that can be called by the latter terms, the only mineral hydroxides being göthite and lepidocrocite, both having the composition $FeO(OH)$.

Ballycastle Coalfield Area.

1" 8. 6" Antrim 3 and 4.

The ironstone of this district is of the "blackband" type, that is, a carbonate mixed with a certain proportion of hydrocarbons, apparently of the same character as those present in oil-shales. This is said to have sufficed for calcination, provided that the combustion was started with a certain admixture of coal. The seam of ore at **Carrickmore**, some three miles east of Ballycastle, where it has been worked by Messrs. Merry and Cunningham, occurs in the Lower Carboniferous Series, some 270 feet (45 fathoms) below a guiding bed of limestone, and 400 feet (65 to 70 fathoms) below the Main Coal. The section shows :—

Roof of blue mudstone.

Coal	6½ ins.
Fireclay	5 ,,
Coal	7 ,,
Ironstone	1 ft. 3 ins.
Fireclay	

The bed must underlie the whole area north of the great east-and-west fault known as the Great Gaw, from Carrickmore West to Bath Lodge. In its eastward extension beneath the southern slopes of Fair Head it apparently undergoes lateral variation, being represented by coal (the White Mine Seam), without any ironstone, where it emerges on the east side of the Head.

Balls of clay-ironstone occur at various horizons, and a seam known as McGildowney's Ironstone is found some 20 fathoms above the Main Coal. This was partially worked in **Gobb** by Messrs. Merry and Cunningham, and is in places 9 inches to 1 foot thick, and similar to the blackband ironstone above described. Min. Stat. 1891, under Ballyvoy, quote it as yielding in that year 625 tons. At the best, only a few inches of coal are associated with it.

I owe much of the foregoing information to Mr. W. B. Wright, who has been engaged on a map and memoir of the Ballycastle coalfield for the Geological Survey of Ireland.

Mr. W. H. Pearson, of Suffolk House, Cannon Street, London, allows us to publish the following partial analysis made by him in 1918 of the blackband ironstone of the 15-inch seam at Sronbane, Ballycastle :—

Iron	30·24
Insoluble matter (chiefly Silica)	12·90
Sulphur	0·02
Phosphorus	0·142
Volatile matter (chiefly hydrocarbons)	33·80
Total loss on calcination	40·30

J. Holdsworth ("Geology etc. of Ireland," p. 60, 1857) notes that some cargoes of "valuable blackband ironstone ... have recently arrived at Glasgow" from Ballycastle, and the first record in Min. Stat. of the use of this ore is the export of 2,000 tons from the "Belfast District" in 1857. In 1858 Ballycastle is definitely mentioned, and it is noteworthy that at that date no other iron ore, not even those in the basaltic series of County Antrim, was being mined in Ireland. The output, which was 3,000 tons in 1859, fell to 46 tons in 1860, with a value of £18 ; but a few years later the annual figure was over 17,000 tons, and rose to 30,000 tons in 1872. Little was done after this, though 1,051 tons were raised in 1880, after which the name of Ballycastle drops out of the official record. "Carbonate ore, County Antrim," is, however, mentioned in 1890 (13,336 tons), and as lately as 1907, when 65 tons are quoted, yielding 20 tons of iron. This material may represent the removal of stocks raised at Ballycastle in previous years.

The Laterite Ores of North-eastern Ireland.

These bedded ores, resulting from the decay in Oligocene times of the surface of the basaltic lavas, have been described in the Memoir of the Geological Survey on the Interbasaltic Rocks (Iron Ores and Bauxites) of North-east Ireland (1912), with the exception of those in Rathlin Island. In 1919 the deep boring put down by the Ministry of Munitions at Washing Bay, on the west side of Lough Neagh, added greatly to our knowledge of the Lough Neagh Clays, which must now be regarded as a product of the denudation of the surface in early Cainozoic times. Evidence is now gathering that there are in some places important lateritic zones on at least two distinct horizons in the basaltic series, as was first recognised in Rathlin Island.

References to the many papers published on these ores will be found in the Memoir above cited, and a good general account is given by J. D. Kendall in his "Iron Ores of Great Britain and Ireland," p. 255 (1893). Special mention should be made of the paper by R. Tate and J. S. Holden, "On the iron-ores associated with the basalts of the north-east of Ireland," Quart. Journ. Geol. Soc. London, vol. 26, p. 151 (1870). Prof. E. Hull (Rep. Select Comm. on Industries (Ireland), p. 75, 1885) estimated the area over which pisolitic iron ore extended in County Antrim as 167 square miles, and the available quantity at that date as 185,500,000 tons. A recent review of the mining area is given in the Report of the Advisory Council on sources of iron, Depart. Sci. and Indust. Research, ed. 2, p. 32 (1918).

While the pisolitic varieties, which are very limited in thickness, graduate into haematite, the beds, as a whole, are laterites, with a considerable percentage of residual silica, and often with a high content of titanium dioxide, commonly as much as 5 per cent. The beds poor in iron graduate into bauxite, while some are mere ferruginous clays, of the "lithomarge" type that results from the decay of basalt under ordinary sub-aerial influences. The relative percentages of alumina and silica, as compared with those in the unweathered basalt, show, however, that the typical ores of the interbasaltic zones have arisen under the conditions of tropical climate and rainfall that give rise to laterite at the present time in other lands.

The ironstone nodules on the surface of the Lough Neagh clays, on the eastern side of the lake, described by Hardman (R.G.S.I., vol. 3, p. 157, 1873) and others, as embedded in these clays, seem to be derived from the destruction of some interbasaltic zone in the neighbourhood; but at some points similar nodules have been found in the main body of the clays.

Kinahan (p. 67) says that attention was first called to the red beds of the north as iron ores in 1861. The first official mention of their working is in Min. Stat. for 1862 (p. 47), where the information given is very inexact, but where it is stated that about 6,000 tons of ore were already going to Ardrossan.

The total of the interbasaltic iron ore raised in north-east Ireland rose to 228,000 tons in 1880, and was nearly 190,000 tons in 1882, and for several years later it represented a value of at least £20,000 per annum. In 1913, the output was 60,014 tons, value £11,290, and in 1915, 39,326 tons, value £7,892, and with an average iron-content of 38·84 per cent.

The names of the mines worked from year to year are given in Min. Stat. from 1869 onwards, when the Antrim Iron Ore Company was stated to be working at Larne, Glenarm, Carnlough, and Glenariff (misprinted Glengariff). In 1870 we read of Ardclinis (printed as Ardshins), Carnlough (printed as Camlough), Glenarm, Kilwaughter, and Glenravel, the last raising 15,000 tons of ore out of a total of 34,582 tons. A good beginning had thus been made.

The mines cited in various years include the following, which are here grouped for convenience in alphabetical order. The names assigned from time to time to the mines in Min. Stat., whether as variants or as mere misprints, are given in italics to avoid future confusion. The references to "Mem." in this list are to the special Memoir on the Interbasaltic Rocks, published in 1912. The maps in that Memoir will aid in identifying most of the mines, and the more important ones are shown by coloured dots on the $\frac{1}{10}$″ Mineral Map. The references to the 1″ sheets in the following list of mines will serve as guides to the older Geological Survey Memoirs, which may be consulted with advantage. Sheets 7 and 8 are dealt with in a combined Memoir (1888). There is a single Memoir for sheets 21, 28 and part of 29 (1876).

Agnew's Hill. 1″ 20. 6″ Antrim 40 N.W. Mem., p. 33.

Ardclinis (*Ardshins*). 1″ 14. 6″ Antrim 20 S.E. On the scarp between Garron Point and Red Bay. Mem., p. 56.

Ballyboley. This name occurs in Min. Stat. 1880, and, no doubt, refers to the openings described in Mem., p. 33, where "Ballyclare Junction" seems to be a slip for "Ballyboley

IRON

Junction." The locality is on the scarp south of Shane's Hill, in 1" 20, 6" Antrim 40 S.W.

Ballylagan (*Ballylaggin, Ballylagon*). 1" 7. 6" Londonderry 3 S.E. Mem., p. 20. The name first appears in Min. Stat. 1878. The mine is in the townland of South Ballylagan, south of Portrush, and north of the adit in Killygreen. It is given in the List of Mines for 1918 as suspended.

Ballypalady. 1" 28. 6" Antrim 51 N.W. and S.W. Mem., p. 91, and Map 3 in Memoir. First quoted in Min. Stat. 1874. Iron ore has been mined under two outliers of Upper Basalt in several townlands in the parish of Ballymartin, west of Ballypalady, and the locality is famous for the fossil plants here found in the interbasaltic zone. Now that the railway station of Ballypalady, which seems to have given its name to the Ballymartin deposits, is known as Doagh, **Ballymartin** would be a better general title for the locality of the mines.

Barard. See Parkmore.

Bay Mines. 1" 14. 6" Antrim 20 S.W. Mem., p. 57. These mines took their name from Red Bay, and are in the townland of Drumnacur.

Broughshane. See Clonetrace.

Cargan. 1" 14. 6" Antrim 24 S.W. Mem., p. 60. Marked by a coloured spot on the Mineral Map. Section in Home Office (No. 1480) showing fault at end of adit, which stopped the working (1883).

Carnlough. 1" 14. 6" Antrim 25 S.E. Mem., p. 78.

Clonetrace. 1" 20. 6" Antrim 28 S.W. Mem., p. 71. N.N.E. of Broughshane, in the south of Clonetrace townland, on the Artoges River. Working in 1918. See also Elginny.

Cloghcor. Under the heading "Glenariff (Cloughcor)," Min. Stat. records an output of 17,857 tons of pisolitic iron ore for 1877. The townland of Cloghcor is in 1" 14, 6" Antrim 24 N.E. and S.E. The mines here, east of the ravine of the Inver River in Glenariff, were active prior to 1886 (see Mem., p. 59).

Correen (*Coreen, Coveen*). 1" 20. 6" Antrim 33 N.W. Mem., p. 69. Immediately south-east of Knockboy Hill. Plan and section, showing beds under basalt, in Home Office, No. 598 (1876). See also Knockboy.

Craigahulliar. 1" 7. 6" Antrim 6 N.E. Mem., p. 20. The great quarry here in columnar basalt has proved far more important than the trials for iron ore or the lignite mine immediately to the south. Craigahulliar is given as a lignite mine in the List of Mines for 1918.

Crommelin. 1″ 14. 6″ Antrim 24 S.W. Mem., p. 46. The mine derives its name from Newtown Crommelin. Working in 1918. See also Tuftarney.

Crossreagh. 1″ 7. 6″ Londonderry 3 N.E. Mem., p. 20. (here misprinted Cropreagh). North of Islandmore, on the road south of Portrush.

Cullaleen. 1″ 19. 6″ Antrim 27 N.W. Mem., p. 82. See Duneany.

Duneany. 1″ 19. 6″ Antrim 27 N.W. Mem., p. 82. This and the Cullaleen adit are under the outlier of Upper Basalt that forms Black Hill, north-west of Ballymena.

Dungonnell. 1″ 14. 6″ Antrim 24 S.E. Mem., p. 61. S.E. of the Cargan mine, and at the head of the Ballsallagh Water. Plan in Home Office (No. 2544) as abandoned, 1891.

Dunluce. 1″ 7. 6″ Antrim 2 S.E. Mem., p. 21. Min. Stat. 1910 record an output; but the mine was not working when visited in 1918.

Elginny (*Elginney, Elgenany*). 1″ 20. 6″ Antrim 28 S.W. Mem., p. 70. North of Broughshane; connected with the Clonetrace mine, which lies eastward from it.

Essathohan. 1″ 14. 6″ Antrim 24 N.E. Marked as a bauxite (A1) mine on the Mineral Map; but Min. Stat. 1900 record an output of iron ore. The List of Mines for 1918 gives the product as bauxite only.

Evishacrow (*Evishnacrow*). 1″ 14. 6″ Antrim 24 N.E. Mem., p. 52. In the Parkmore area. Working in 1918.

Evishnablay (*Evishnably*). 1″ 20 (extreme north). 6″ Antrim 28 N.W. Mem., p. 64. One of the Mountcashel mines, on the outcrop east of Gortnageeragh village.

Glebe. 1″ 20. 6″ Antrim 29 N.E. Mem., p. 29. In the townland of Glore, 1½ miles S.W. of Glenarm, and a little south of the coloured spot on the Mineral Map. First quoted in Min. Stat. 1875, and working in 1907. See also Rep. Advis. Council on sources of iron, 1918, p. 33.

Glenariff. This name in Min. Stat. probably includes the Bay Mines (Mem., p. 57) and the Glenariff Mines (Mem., p. 59). See also Cloghcor.

Glenarm. 1″ 20. 6″ Antrim 29 N.E. Mem., p. 28, Map 5 in Memoir. Old workings occur at Fallmore village, north of the Glebe mine. Coloured spot on west side of Glenarm River on the Mineral Map.

Glenravel Mines. 1″ 14. 6″ Antrim 24 N.W. Mem., p. 43. Associated with "Evishacrow, No. 3," in List of Mines for 1918.

Gortnageeragh. 1″ 20 (extreme north). 6″ Antrim 28 N.W. Mem., p. 64. One of the Mountcashel mines, in Gortnageeragh townland, S.E. of the village. Coloured spot east of Cross Roads station on the Mineral Map. See also Evishnablay.

Irish Hill and **Straid.** 1" 28. 6" Antrim 46 S.W and 52 N.W. The mines are in the outcrop below an outlier of Upper Basalt that forms Straid Hill in its northern and Irish Hill in its southern part. Mem., p. 36. First quoted in Min Stat. 1875. Bauxite is produced as well as iron ore.

Island Magee. 1" 21. 6" Antrim 41 N.W. Mem., p. 34. The mine is in Ballylumford, and is quoted in Min. Stat. 1880 for the first time. The 6" map of this district has been published geologically coloured.

Islandmore. 1" 7. 6" Londonderry 3 N.E. The townlands of Upper and Lower Islandmore are in the N.E. Liberties of Coleraine. The name appears in Min. Stat. 1902, with a small output of 150 tons of ore, averaging 35 per cent. of iron. This was increased to 2,000 tons in 1905, but sank to 120 (iron 28 per cent.) in 1907, after which date no special quotation is recorded.

Killygreen. 1" 7. 6" Londonderry 3 S.E. Mem., p. 19. Immediately south of Islandmore. Returned as producing iron ore in Min. Stat. 1914, but was mainly a bauxite mine. The adit was closed when visited in 1918.

Kilwaughter. 1" 20. 6" Antrim 40 N.W. Mem., p. 33. Quoted as producing " haematite " in Min. Stat. as far back as 1871 ; later given as " brown haematite." This mine is about two miles S. by E. of Agnew's Hill.

Knockboy. 1" 20. 6" Antrim 33 N.W. One of the Correen Mines under Knockboy Hill and S.W. of Correen village. Mem., p. 69. Appears first in Min. Stat. 1874. Home Office plan (No. 1163) gives it as abandoned, 1880.

Mains (also written *Main*). This name appears as that of a mine near Portrush in Min. Stat. 1879, and is recorded under the name of Main in 1880 and 1881 The ore is said to have been aluminous.

Mountcashel. The name of a group of mines including Evishnablay and Gortnageeragh. Mem., p. 64. *Mount Cashel* is given as working in the Home Office List of Mines for 1918.

Newtown Crommelin. See Crommelin. Mem., p. 46.

Parkmore. 1" 14. 6" Antrim 24 N.E. The mines are at the head of the Ballymena and Parkmore railway, where Fe is marked on the Mineral Map.

Rathkenny. 1" 20. 6" Antrim 28 S.W. About half a mile east of Rathkenny station on the Ballymena and Parkmore line. Mem., p. 66. Working in 1918.

Rathlin Island. 1" 8. 6" Antrim 1. Memoir 7 and 8, p. 27.

Red Bay. An analysis is given under this name in Rep. Advis. Council on sources of iron, 1918, p. 33. See Bay.

Red Rock. This name appears as near Larne in Min. Stat. 1872, with an output of 3,468 tons of ore.

Shane's Hill. 1″ 20. 6″ Antrim 40 S.W. Mem., p. 33. The output from mines on the scarp at Shane's Hill is given in Min. Stat. for 1871, 1872 and 1873. The Report of the Advisory Council on sources of iron (1918, p. 33) records the abandonment of working here in 1886.

Slievenanee (*Slieve an Nee*). Min. Stat. 1881 record this name in connexion with the Glenravel workings. Mem., pp. 42 and 55. An analysis is given under the name *Slievananee* in Rep. Advis. Council on sources of iron, 1918, p, 33. showing 65·2 per cent. of metallic iron.

Solomon's Drift. One of the Crommelin mines (see Crommelin) on the east side of the Skerry Water valley. Mem., pp. 48 to 50. Mainly worked for bauxite.

Straid. See Irish Hill.

Trostan. This name occurs in Min. Stat. 1872 and 1873, and, no doubt, refers to a mine on the scarp north of Parkmore station (1″ 14. 6″ Antrim 19 and 24; Mem., pp. 42 and 55). Plan in Home Office (No. 149) as abandoned, 1874.

Tuftarney. 1″ 14. 6″ Antrim 24 S.W. Min. Stat. 1914 give an output of iron ore; but the mine is mainly worked for bauxite. It is on Tuftarney Hill, an outlier of Upper Basalt, about one mile east of Newtown Crommelin. Mem., p. 50.

Urbalreagh (*Urblereagh, Urblereigh, Orblereigh*). 1″ 7. 6″ Antrim 6 N.E. The mine is in the fork of the two roads running southward near Ballymacrea House. The first record is in Min. Stat. 1877. J. D. Kendall ("Iron·ores of Great Britain and Ireland," p. 259, 1893) gives a section showing 1 foot 9 inches of pisolitic ore and 7 feet of bole.

Cookstown Area.

The area between Cookstown and Lissan (1″27. 6″ Tyrone 29 N.E. and S.E.) contains several old workings in veins and layers of nodular limonitic ore, occurring in Lower Carboniferous strata. These were mainly in the adjacent townlands of **Unagh** and **Tullycall** (6″ 29 S.E.). Unagh is on the high road running from Cookstown over the western shoulder of Slieve Gallion, and is only two miles from the centre of the town. Griffith (1861, p. 152) places it on Slieve Gallion (see below), in which slip he is followed by Kinahan. Boate (1652, p. 128) knew of " white-mine " ore, possibly carbonate, " at the foot of the mountains Slew-galen." The district is described by E. T. Hardman, " On the occurrences of siliceous nodular brown haematite (göthite) in the Carboniferous Limestone beds near Cookstown," R.G.S.I., vol. 3, p. 150 (1873). A section of the beds is given, and the author supports his determination of the ore as göthite by the following analysis (p. 153):—

Fe_2O_3	74·56
FeO	a trace
MgO	0·044
CaO	a trace
SiO_2, etc.	9·42
Al_2O_3	3·51
Water combined	10·24
Water hygroscopic	2·90
MnO_2	a trace
CO_2	a trace
P_2O_5	a trace
	100·674

Metallic iron 52·20.

Hardman is responsible for the name *Cove Bridge*, probably derived from Cove Kilns, under which the mines are mentioned in Rep. Advisory Council on sources of iron, p. 35 (1918).

The Barrow Haematite Steel Company made searches in this area (townlands of Unagh, Tullycall, Toberlane and Coolreaghs) in 1875, in shafts varying in depth from 90 to 172 feet, and they are recorded as raising 572 tons of brown haematite in 1880 (Min. Stat.), value £286. Plans were, however, deposited in the Home Office as of abandoned mines on 25th January, 1881. The district appears still to offer considerable possibilities.

Slieve Gallion (Sliabh-Gallan, including **Carndaisy** and **Tirgan).** 1" 27. 6" Londonderry 46 N.W. Quartz veins with haematite traverse the granitic and doleritic rocks that form the main mass of Slieve Gallion, north-west of Moneymore. They probably attracted attention, from their striking colour on the hillside, in prehistoric times. Kinahan (pp. 48 and 92) states that they were worked by Rennie in the seventeenth century; G. V. Sampson (Stat. Surv. Londonderry, p. 104, 1802) says that the mining was done by an agent of the Drapers' Company. As noted above under Unagh, there has been some confusion of the veins in the Carboniferous strata in the townland of Unagh with those occurring on Slieve Gallion. The mountain is described by G. A. J. Cole, " On the Geology of Slieve Gallion," Trans. R. Dublin Soc., vol. 6, p. 213 (1898), where references to earlier literature will be found. In the Report of the Advisory Council on sources of iron, already referred to, p. 36, (1918), an analysis, communicated by W. S. Gresley, gives 61·30 per cent. of iron in the ore. The admixture of quartz in the veins is seen to vary greatly in the field. A scheme for working this ore, among others, was put forward in Belfast in 1907, and the prospectus ("Derry and Antrim Ore Co.") states that a trial shaft was put down.

Magheramenagh. 1″ 32. 6″ Fermanagh 8 N.E. Haematitic ore is said to have been associated with copper ore in the north part of this townland, which lies east of Belleek. A plan and section in the Home Office series (No. 397), under the name **Belleek,** shows the iron ore lying horizontally in the Carboniferous Limestone, thinning out eastward, and about 50 feet from the surface. Worked by the Wigan Coal and Iron Company, and abandoned in 1875.

Castle Caldwell. 1″ 32. 6″ Fermanagh 9 N.W. Haematitic ore occurs in veins in Rossbeg, associated with copper ore (Griffith 1861, p. 145). The sites of explorations made in 1877 are shown in Plan 658 in the Home Office series. The MS. 6″ Geological Survey map, Fermanagh 8 N.E., records haematite in the Carboniferous Limestone of Leggs townland, west of Rossbeg. See Magheramenagh above, which lies a little farther west.

Tyrone Coalfield. 1″ 35.

A certain amount of nodular clay-ironstone exists in connexion with the coal-bearing strata of **Coal Island** and **Dungannon.** It was known to Boate (p. 128, 1652). E. T. Hardman (Mem. 35, p. 87, 1877) gives determinations of the metallic iron content of nodules from four horizons, the figure varying from 21·7 to 35·5 per cent. Kinahan (p. 42) suggests that the drift conceals much of the iron-bearing strata.

Deehommed (sometimes pronounced "Dechomet"). 1″ 48. 6″ Down 35 N.E. Haematite occurs in the altered Silurian slates, near the margin of the granite, $\frac{1}{4}$ mile west of the hamlet of Bing's End, and N.E. of Katesbridge station on the Scarva and Newcastle line. The ore is first mentioned in Min. Stat. 1874, when 3,000 tons, valued at £1,800, were raised, but apparently not put upon the market. No return is given for 1875. The spot has been often visited by prospectors, and was reported on by Mr. H. J. Daly to the Department for the development of Mineral Resources in 1917. The extent of the ore cannot be estimated until systematic opening-up has been undertaken.

Glaslough. 1″ 46. 6″ Monaghan 7 N.W. and S.W. Coote (Stat. Surv. Monaghan, p. 151, 1801) records ironstone as raised here on the bank of the Knockbawn millstream. I can trace no other record.

Ballynakill. 1″ 55. 6″ Sligo 27 S.E. East of Srananagh church, and just north of the main road from Ballygawley to Ballyfarnan, haematite occurs in Lower Carboniferous strata, in the south of Ballynakill townland. There are considerable traces of old furnaces immediately to the west. It is quite likely that this is the *Ballonakill* known to Boate (p. 130, 1652) as an iron-furnace worked by the Earl of Londonderry early in the 17th century. Kinahan (p. 73) states that carbonate ores were brought here from the hills to the east for smelting.

Cuilcagh. 1″ 56. 6″ Cavan 6 N.E. Sir C. Coote (Stat. Surv. Cavan, p. 24, 1802) says that the iron furnaces and forging mills round which Swanlinbar arose were fed by ore from " Quilca " Mountain. This was possibly bog ore, in whole or in part. Coote traces the name Swanlinbar to the names of the four workers of the iron mine, Swift, Saunders, Darling, and Barry. He writes the name " Swanlingbar." A similar way of forming names is now common in commercial circles in N. America.

Lough Allen Coalfield. 1″ 55, 56, and 67.

The beds of iron carbonate associated with the Upper Carboniferous strata in this district have often been described. See especially Griffith, " Geological and Mining Survey of the Connaught Coal District," 1818; also Kane 1845, p. 133. The ore has been mined at various points in the east of the area of 1″ 55, as shown near Drumkeeran and Belhavel on the Mineral Map, and a little farther north in the area of 1″ 56. The **Creevelea** Iron Works (1″ 55. 6″ Leitrim 16 S.W.) are mentioned in Min. Stat. 1857, pp. 57 and 91, under the name *Creveela*. J. Kincaid (" A section across the Coal beds of Leitrim, " G. S. D., Vol. 7. p. 301, 1857) refers to their working at this date, and a good historical account, with references, is given in Mem. 55, pp. 29-30 (1885). See also Kin., p. 73. The abandoned works still give an impression of the former mineral activity of the district. The ironstone nodules are well seen in the floor of the valley of the **Arigna** River (1″ 67. 6″ Leitrim 20 S.W. and 20A N.E.; also 6″ Sligo 35 N.E. and Roscommon 2), and the position of the layers is shown in the Geol. Surv. Longitudinal Section 33 (1891). The Geological Survey has published coloured maps of this coalfield on the 6″ scale. **Slieve Anierin** (Irish Sliabh-an-iairn), the

Iron Mountain, east of Lough Allen, derives its name from these carbonate ores, and was known to Boate ("Ireland's Naturall History," p. 128, 1652) as a place, *Slew Naren*, where ore was dug before 1645. The O'Reilly brothers used this ore with that of Arigna about 1788 (McParlan, Stat. Surv. Leitrim, p. 12, 1802).

Smelting was carried on at Drumshambo down to 1765 (Griffith, "Geol. and Mining Survey of the Connaught coal district," p. 57, 1818), and in the Arigna Valley, near Lough Allen head, from 1788 to 1808, and from 1824 to near the middle of the 19th century. Isaac Weld (Stat. Surv. Roscommon, pp. 36-50 and Appendix, 1832) goes very fully into the history and prospects of the iron works, and says that an English staff and miners were employed from 1824-6, and that some revival was attempted in 1831. In those seven years, however, only 300 tons of iron had been produced at a cost of £50,000. The details of a Parliamentary enquiry into the position of the Arigna Iron and Coal Company, the report of which was printed in 1826-7, are discussed in Weld's appendix. Kinahan summarises the whole history (pp. 72 and 73).

The older works at Slieve Anierin used wood charcoal; but, as Griffith points out (op. cit., p. 59), the O'Reillys used the local coal, and were in this matter the first pioneers in Ireland.

P. Buchan, in a somewhat discursive paper entitled "On the composition of the Iron Ores of the Connaught Coalfield" (Journal R. Dublin Soc., vol. 2, p. 1, 1858) gives a general history of the working of the Irish ores, and furnishes numerous analyses of the carbonate nodules of the Lough Allen district, especially round Creevelea. These show 47·65 to 82·07 per cent. of iron carbonate. Kane (1845, pp. 133-167) considered at length the smelting of iron from the Irish carbonate ores, and his analyses and high opinion of samples from Arigna (p. 135) have been often quoted, an average giving 40 per cent. of iron.

A plan of the ironstone workings at Arigna, dated 1812, is in the Home Office series (R 284).

Calliagh. 1" 58. 6" Monaghan 13 S.W. Memoir 58, edition of 1914, p. 18, describes the band of ferruginous shales and iron ore at Logwood Hill in Calliagh, about 5 miles S.S.W. of Monaghan town. W. E. Adeney made several analyses from the ore in the quarry on the hill, which were printed in a Report made to Colonel Lloyd in 1888. The richest ore gave, in its soluble portion, ferric oxide 42·2 and manganese peroxide 6·24 per cent., with only 0·03 of phosphoric oxide; but the insoluble residue was 37 per cent. of the ore. This analysis is quoted by J. O'Reilly, "On idocrase in County Monaghan,"

Proc. R.I.A., ser. 3, vol. 1, p. 446 (1889-91). Dr. Adeney's report was unfavourable, owing to the presence of a very high percentage of silica. Kinahan, p. 480, refers to the locality as "Calliagh and Tattin Heive;" the latter should be Tattintlieve. See also under Manganese.

Mullenmore North. 1" 64. 6" Mayo 38 N.E. J. McParlan (Stat. Surv. Mayo, pp. 19-20, 1802) observed "beautiful red iron ore" in heaps round Rutledge's ruined iron-furnaces at *Mallinmore*, which he states is on a branch of the Deel Water. This is evidently the townland of Mullenmore North, where the 6" MS. Geological Survey map records " Old furnace. Red haematite," near a small stream, unconnected with the Deel, running eastward to Lough Conn. The source of the ore is said by McParlan to have been " Deel Mountain "; it would be interesting if it could be traced. Mem., 41, 53, 64 (1879) makes no mention, though its author records the ore-heaps on the ground.

Redhills (Claragh). 1" 68. 6" Cavan 16 N.W. The mine here was on Ordovician slates, north of Claragh lake, and was actively working in 1872. E. Hull, in 1872, described the beds in a brief paper in Journ. Roy. Dublin Soc., vol. 6, p. 216 (1875), and a fuller account is given in Mem. 68 and 69, p. 22 (1873). The ore is limonitic. An analysis by John Cameron is quoted as follows :—

Peroxide of Iron	57·57
Peroxide of Manganese	traces
Protoxide of Manganese	6·20
Alumina	8·93
Carbonate of Lime	0·50
Silica	22·80
Water of combination	3·00
Soluble matter	1·00
	100·00

Metallic iron 40·30 per cent.

Carrignahelty. 1" 73. 6" Mayo 75 N.E. It seems worth while to call attention to the occurrence of haematite in a vein " of extraordinary richness in Precambrian quartzite," recorded by W. F. Mitchell in Mem., 62, 73, p. 21 (1879). The spot is in the townland of Bolinglanna, and is marked with an iron sign on the 1" geological map. The MS. 6" sheet records " trial workings for red haematite," and the ore is associated by Mitchell with the occurrence of iron pyrites in the district (see Gubnabinniaboy, under Sulphur). The site is below the name *Corraun* on the Mineral Map.

Gortinee. 1" 78. 6" Leitrim 35 N.W. This is an occurrence of limonitic ore in an inlier of Ordovician shale, east of and close to the railway from Longford to Carrick-on-Shannon, and 3 miles S.E. of Drumsna. See Mem. 78, 79, and 80, p. 41 (1873). Griffith (1861, p. 148, and Map) mentions the iron mine here, and Kinahan (p. 90) states that ore was raised from it as far back as the 17th century. P. Buchan ("Iron ores of the Connaught Coalfield," Journ. Roy. Dublin Soc., vol. 2, p. 20, 1858) refers to the locality under the name of **Derrycarne** and says that the ore was smelted at Dromod as late as 1798. Derrycarne Demesne is a townland 1 mile to the south, and Mr Nesbitt, its owner, worked the Gortinee ore in the 19th century (Kin., p. 90, says between 1860 and 1880), shipping it to Wales. Previously, in the same century, it had been mixed with the Arigna carbonate ores by Messrs. Reilly and the Arigna Company.

Cleenrah (*Cleenragh*) and **Enaghan.** 1″ 79. The latter and more northern locality is in 6″ Longford 3 N.W. ; Cleenrah is 2 miles to the south in 3 S.W. and 6 N.W. The two coloured spots appear under the one name Cleenrah in the Mineral Map. At Cleenrah, on the west side of Lough Gowna, three beds of limonitic ore occur in Ordovician shales. They were worked as recently as 1873 ; but Min. Stat. record no output. See Mem. 78, 79, and 80, p. 41 (1873), and E. Hull, paper cited under Redhills above.

Kilbride. 1″ 120. 6″ Wicklow 5 N.E. and 6 N.W. Excavations were made E.S.E. of the village of Kilbride, near Blessington, in the townlands of **Knockatillane** and **Cloghleagh,** on a vein running N.W. on the hillside of the Lackan hollow. See Griffith Map (1855) and the 1″ geological sheet; also Mem. 120, p. 18 (1880). In the Min. Stat. List of Mines, Cloghleagh (*Cloughleagh*) appears as worked by A. Illingworth, 1862-8, and Knockatillane by the Dean of Clogher, from 1862-6.

The occurrence is described by S. Haughton ("Geological notes on the Iron and Manganese ores, and on the China Clay of Kilbride," G.S.D., vol. 10, p. 69, 1864). The lode is stated to contain " brown haematite " and psilomelane, and an analysis is given of the " pitchy iron ore " as follows :—

Peroxide of Iron	77·15
Water	20·43
Phosphoric acid [P_2O_5]	1·60
Argil [insoluble residue]	1·30
	[100·48]

In the notes furnished by the Geological Survey to the Report of the Advisory Council on sources of iron, p. 35 (1918), Haughton's analysis appears by an error under the heading of Ballard, which is near the eastern Kilbride in the same county.

Kildavin. 1″ 124. 6″ Clare 20 S.W. Limonitic ore was raised here in the old days of local smelting from the Silurian rocks on the south slope of the Graney valley, 2 miles N.E. of Feakle. There is still a hamlet called Furnace near Feakle. The district also produces bog iron ore. Mem. 124 and 125, p. 47 (1863) and Kin. pp. 44 and 75. The adit mentioned in the Memoir in Glendree, some 5 miles to the west, was probably for lead ore ; see under Lead.

Woodford. 1″ 125. 6″ Galway 125 S.E. The old furnaces here were supplied in part from local bog ore. See under Bog iron ore.

Tomgraney. 1″ 134. 6″ Clare 28 S.E. Similar ore to that of Kildavin was formerly raised here from the Silurian shales and was smelted locally or on the shores of the waterway of Lough Derg. Mem. 134, p. 43 (1861), names mines in **Ballymalone** and **Bealkelly.** (6″ Clare 29 S.W.). The MS. 6″ geological map shows the sites in these adjacent townlands south of Scariff Bay.

Leinster Coalfield.

Nodular carbonate ores have been mined at various points from beds in this area, but have not assumed the importance of those in the Lough Allen (Connaught) field. Griffith ("Report on the Leinster Coal district," p. 25, 1814), refers in a general way to ancient excavations in a bed of nodular ore. This (*ibid.*, p. 19) was 32 feet below the lowest coal. In his records of borings, Griffith gives information as to numerous thin seams of ironstone in the Coal-Measure strata. Kane (1845, p. 131-2), by adding up their thicknesses, arrives at a respectable result ; but the reader would hardly infer from this that the seams were usually one to two inches thick, the best being 4 inches. Kane quotes two analyses, yielding respectively 39·7 and 37·6 per cent. of metallic iron.

Cullenagh Hill. 1″ 127. 6″ Queen's County 24 N.W. and N.E. The carbonate ore was worked on the outcrop of the slope east of Ballyroan, below the lowest coal-seam. See Mem. 127, 128, etc., p. 29 (1881) and Kin., p. 97. For its use in the 17th century at Mountrath iron-works, see below under Dysart.

Aghamucky. 1″ 137. 6″ Kilkenny 6 S.W. W. Tighe (Stat. Surv. Kilkenny, p. 73, 1802) says that nodular iron ore, evidently siderite, occurs as a bed 13 to 15 inches thick in Aghamucky, east of Castlecomer. He believed that the Mountrath works, famous in the 17th century (see under Dysart), had been supplied from this district. The Geological Survey 6″ sheet does not record the ironstone nodules, and Griffith's section ("Report on the Leinster Coal district," p. 106, 1814) in "*Ahamucky*" names no ironstone in a depth of over 100 feet.

Dysart. 1″ 128. 6″ Queen's County 13 S.E. This mine was worked in a deposit of limonitic ore in Carboniferous Limestone, close to and east of Dysart old church. Boate ("Ireland's Naturall History," p. 127, 1652) classed this brittle ore as Rock-mine, and records it as found only near Tallo, in Munster, and "in King's County" (apparently in error) " in a place called Desert land, belonging to one Sergeant Major Piggot." (Compare Kinahan, p. 90). Before 1641, it " furnished divers great Iron-works," though the pits were only 1 fathom deep. Boate seems to have regarded the mineral as a kind of bog-ore, since he refers to the rich soil that overlies it. On p. 135 (p. 75 of the 1755 edition), he shows, by using the term rock-mine, that this ore was used by Sir Charles Coote at his extensive works near Mountrath (6″ Queen's County 17 N.W.). One part of it was mixed with two parts of white-mine (carbonate ore), which came " from a place two miles further off." Probably this was the nodular ore of Cullenagh Hill; both Dysart and Cullenagh are about 10 miles from Mountrath town. See also Aghamucky. The iron produced was exported through Waterford, and in Mem. 127, p. 28, it is stated that no trace of mines has been found near Mountrath itself. The ore of Dysart was analysed by Jas. Apjohn for Lord Carew (Mem. 128, p. 30, 1859), and yielded 34 to 36 per cent. of iron. Holdsworth (" Geology etc. of Ireland," p. 59, 1857) speaks of this work, done in 1854, as a new discovery of ore.

Imail (*Imael*). 1″ 129. The Mining Company of Ireland (M.C.I. 1860 ii) leased an iron mine in the Glen of Imail from Lord Wicklow in 1859; but it was reported on as difficult of access, and searches were abandoned in 1860. The glen (6″ Wicklow 21 N.E. and S.E. and 22 N.W. and S.W.) has been carved out in the granite south of Donard by the head-waters of the Slaney. The name is not on the older 1″ sheet, but can be found on that published in 1913.

Ballard (Ballycapple). 1″ 130. 6″ Wicklow 30 S.E. and 31 S.W. This mine has been opened on a remarkable lode of magnetite, six miles south-west of Wicklow town. The lode runs parallel to and about 300 yards north of the road from the hamlet of Ballycapple to Ballard. Shafts and openings have been made in Ballard Upper and adjoining townlands, south-west of the village of Kilbride. C.R.C. Tichborne (" On the occurrence of magnetic oxide of iron at Kilbride, County Wicklow," R.G.S.I., vol. 4, p. 264, 1877), shows by the latitude and longitude given that he is dealing with this lode and not with that at Kilbride near Blessington. He gives two partial analyses of the ore, showing:—

	I.	II.
Magnetic oxide	40·10	72·49
Ferric oxide	22·03	15·30
Oxide of manganese	2·05	trace

The specific gravity of specimen II. was 4·31. Tichborne attributes the formation of the magnetite to thermal influences acting on ferrous sulphate in solution.

Mr. W. H. Pearson, of Suffolk House, Cannon Street, London, kindly allows us to publish the following analysis made by him in 1918 of the manganiferous portion of the ore :—

Peroxide of iron	46·24
Protoxide ,,	1·68
Peroxide of manganese	18·08
Protoxide ,,	6·18
Alumina	5·96
Silica	11·70
Lime	1·40
Magnesia	0·36
Sulphur	0·055
Phosphoric acid [P_2O_5]	0·286
Arsenic	0·134
Copper, Lead, Zinc	nil
Combined water	7·68
	99·755

Iron	33·68
Manganese	16·22
Phosphorus	0·125

Mr. Pearson remarks that the magnetic iron ore is variable in quality, the richest samples yielding about 60 to 64 per cent. iron, with 6 to 11 per cent. silica, and 0·10 phosphorus. He says that " good average ore could be mined there, running about 50 to 55 per cent. iron."

The most detailed description of the Ballard lode hitherto published is that by P. H. Argall ("Ancient and recent mining operations in the East Ovoca district," R.G.S.I., vol. 5, p. 162, 1880). He says that the first workings were made some two hundred (now 240) years ago. Kinahan, p. 112, gives a good history of the sporadic ventures on the lode ; in 1876 Tichborne (*op. cit.* above) stated that 20 tons of ore were raised in 1876. Min. Stat. 1891 names the mine as again working, and it probably accounted for some of the 235 tons of iron ore exported from Wicklow in that year (Custom House report, quoted in a footnote in Min. Stat.). Exploration of the old workings was resumed in 1918.

The lode is no doubt connected with the great mineral belt of the Ovoca mines. It runs approximately in the strike of the Ordovician series, which here includes a number of igneous bands. Examination in the field suggests that it may be a replacement of rhyolite, like the banded ores described by A. E. V. Zealley ("On certain felsitic rocks, hitherto called 'banded ironstones,' in the ancient schists round Gootoma, Rhodesia," Trans. Geol. Soc. S. Africa, vol. 21, p. 43, 1919.)

Thin sections, however, give no clue, and only serve to show a considerable quantity of chalcedonic quartz associated with the granular iron ore. The lode in its upper part becomes limonitic and rich in manganese oxide (see under Manganese).

Glendalough. 1" 130. 6" Wicklow 23 N.W. Smyth ("Mines of Wicklow," cited under Ovoca, p. 358) called attention to a vein of iron carbonate 3 feet wide, accompanied by quartz and a dyke of soft granite, at the head of the valley of Glendalough. The presence of iron carbonate is noted on the MS. 6" map in the office of the Geological Survey. Some operations seem to have been made on this crystalline mass, but no definite mine has been opened up.

Ovoca (Avoca) Mines.

Sir W. W. Smyth's memoir " On the mines of Wicklow and Wexford " (Rec. School of Mines, vol. 1, part 3, 1853) remains the classic account of the important mineral belt in the Ordovician rocks of the county of Wicklow. The mines produce mainly copper and iron pyrites; but the oxidation of the pyritous lodes has allowed some of them to be worked as sources of iron, or of ochre for the colourists' trade.

Connary (*Connoree*) (1" 130. 6" Wicklow 35 N.E.) is mentioned in Min. Stat. 1880 as yielding 20 tons of ochre, and in 1882 and later years as raising some hundreds of tons. **Cronebane** (1" 139. 6" Wicklow 35 N.W. and N.E.) yielded 2,155 tons of " brown haematite " in 1872, 4,143 in 1873, but fell off by 1876. **Tigroney** (1" 139. 6" Wicklow 35 N.W.) is grouped with its neighbour Cronebane as raising 12,180 tons of iron ore, no doubt limonitic ore, in 1883, value £6,394, and also ochre in this and succeeding years, ochre being recorded as 1,105 tons in 1899. **Ballygahan** (1" 139. 6" Wicklow 35 S.W.), in Ballygahan Lower, on the opposite side of the river to Tigroney, produced ochre. **Ballymurtagh** (same sheets), immediately S.W. of Ballygahan, appears in Min. Stat. 1856 with 441 tons of " oxide of iron," no doubt ochreous, and as much as 25,816 tons of " brown haematite " in 1864; but a rapid fall in output then occurred. 16,433 tons are recorded, however, for 1874; but considerable fluctuations took place in the trade, and only 300 tons of ochre are cited for 1883. In recent years, working has been fairly continuous, and the Irish Ochre and Minerals Company raised 1,005 tons of ochre from Ballymurtagh in 1913. In the Home Office List of Mines for 1918 it is given as worked by the Via Gellia Colour Company of Matlock Bath, Derbyshire, and employing 30 persons.

S. Haughton ("Notes on Irish Mines," G.S.D. vol. 5,

p. 281, 1853) gives an analysis of haematite from Ballymurtagh, by J. A. Galbraith, as follows :—

Peroxide of iron	74·37
Clay and Silica	11·00
Water	14·12
Volatile matter	0·41
Loss	0·10
	100·00

Metallic iron 52 per cent.

The term "black band" applied to the ore of Ballymurtagh in Min. Stat. 1869 is probably an error, and in 1871 the ore is styled "aluminious."

A view of the Ovoca ochre-works (Via Gellia Colour Company) is given in E. St. J. Lyburn's paper in Journ. Depart. Agric. and T. I., vol. 16, 1916 (fig. 7).

Kilcashel (1" 139. 6" Wicklow 35 S.W.), a townland between Knockanode and Ballymurtagh, appears as the name of a mine raising iron pyrites and ochre in the Home Office List of Mines for 1917.

Ballymoneen and **Knocknamohill.** 1" 139. 6" Wicklow 35 S.W. and 40 N.W. respectively. Kin. pp. 34, 51, 110, and 118 quotes these as the Knocknamohill Mines. They are described as the West Ovoca Mines in Mem. 138 and 139, p. 31 (1888). The townland of Ballymoneen is S.W. of Ballygahan Upper and of the Ballygahan and Ballymurtagh mines, and its centre is about 1 mile west of the Ovoca River. On the south it adjoins Ballinpark, and the MS. Geological Survey 6" sheet records on this boundary "shafts and levels in a bed of quartz and specular iron, iron pyrites, and chlorite." The ore raised here is said by Kinahan to have been limonite ; but magnetite is now known to occur. Half a mile farther to the east, there is a note on the MS. 6" map of a "level driven in slate."

Knocknamohill townland touches Ballymoneen at its N.W. angle, and extends southward to the bank of the Aughrim River, where the second Fe in the symbol attached to the name Ballymurtagh appears on the Mineral Map. In the N.E. part of the townland the 6" MS. Geological Survey sheet records "ancient open works," and "copper mines here about 1846 (Crockford and Company)." There seems to be no record of output, but iron ore was raised and may still be developed here, as in Ballymoneen. The two mineral occurrences are in the line of the magnetite ores of Ballycoog and Moneyteige, which lie south of the Aughrim River. Kinahan (p. 118) quotes a favourable report on the ore by Dr. W. E. Adeney, and a report was recently made by Mr. Lett for the Department for the development of Mineral Resources. The Department permits us to quote that both in Knocknamohill and Ballymoneen Mr. Lett describes the ore as a mixture of magnetite and haematite, in green chloritic schist.

Moneyteige (*Moneyteigue*). 1″ 139. 6″ Wicklow 39 S.E. The lode, occurring in the north-east corner of Moneyteige South, is magnetite, like that of Ballard, and is similarly connected with the great mineral belt of the Ovoca district. Smyth mentions this occurrence in his " Mines of Wicklow and Wexford " above cited, p. 371. See also Kin. pp. 111 and 118 ; both Kinahan and Griffith (1861, p. 153) state that copper ore is associated with the iron, and Griffith adds " particles of Gold." Kinahan regards the workings as among the oldest in the district.

Ballycoog (*Ballycouge*). 1″ 139. 6″ Wicklow 39 N.E. In the townland of Ballycoog Upper, towards the south bank of the Aughrim River, there is a similar lode, possibly a continuation of that at Moneyteige. A. Mills ("Second Report on the Wicklow Gold Mines," Trans. Dublin Soc., vol. 3, p. 81, 1803) possesses a sketch of a mine here, which was working about 1769. The lode showed magnetite, blende, and copper and iron pyrites. Stewart (p. 121, 1800), whose authority is not convincing, said that it was the richest black iron ore that he knew, and he noted the site of an ancient mine in " Ballycouge." R. Fraser (Stat. Surv. Wicklow, p. 21, 1801) records the old shafts for copper ore and magnetite. The lode is well marked in Smyth's map (" Mines of Wicklow," 1853), and the words " Level made in search of copper " appear on the engraved 1″ topographical map. See also Kinahan, p. 118, who notes the old workings.

Loghill (**Rock Lodge Colliery**). 1″ 142. 6″ Limerick 9 S.E. Mem. 142, p. 40 (1860) records clay ironstone as formerly extensively worked here along the outcrop in Upper Carboniferous strata. Kinahan (p. 42) suggests that this was in the sixteenth and seventeenth centuries.

Glin. 1″ 142. 6″ Limerick 17 N.E. Old workings on clay ironstone, but no indication on the MS. 6″ map. Same references as those given under Loghill.

Shanagolden. 1″ 142. 6″ Limerick 19 N.W. Mem. 142, p. 40 (1860) records " siliceous haematite " in beds of ochreous clay in the stream-bank ¼ mile south of the village. There was no mine. The occurrence is in Ballycormick, and the MS. 6″ map records " red iron ore in yellow clay, same as Kilcoman."

Kilcolman. 1″ 152. 6″ Limerick 28 N.W. Kinahan (p. 47) says that " siliceous limonite " was mined here about the seventeenth century and about 1870-75. Mem. 152, p. 27 (1860), speaks of " siliceous haematite at an old mine S.

of the village "; the ore was examined by A. Gages and found to be too rich in silica. The MS. 6" map of the Geological Survey records this " old sinking for iron " in the townland of **Knockbweeheen,** and notes " red iron ore." Griffith has no reference.

Araglin. 1" 166 (extreme S.W. of Sheet). 6" Cork 28 N.E. and 28 A. The village of this name is not shown on the old engraved sheet used for the map of the Geological Survey, but it will be found on the present topographical sheet. It lies in the county of Cork, at the junction with the counties of Tipperary and Waterford. There are traditions of mines near this spot, and extensive smelting of iron ore was carried on at furnaces south of the Araglin River, in County Waterford, by Lord Cork in the seventeenth century (Mem. 200, etc., p. 26). Boate ("Ireland's Naturall History," p. 137, 1652) states that " knowing persons, who have had a particular insight into his affaires," held that Lord Cork had profited above £100,000 clear gain by his iron-works in various parts of Munster (see also *ibid.*, p. 129). The chief of these works seems to have been at Araglin, and the ore was probably the " rock-mine " or limonite said by Boate (p. 127) to have been raised near Tallow. Tallow lies on the Bride about 10 miles S.E. of Araglin, in 1" 177. It is remarkable that the source of so considerable an amount of ore should be now unknown. Kinahan (pp. 45 and 78) interprets Boate as saying that the smelting works were at Tallow ; but Boate is really perfectly clear. Captain Ablett, who visited Araglin in 1918 on behalf of the Ministry of Munitions, informs me that, like Kinahan, he was unable to find local sources of the ore. No spot can, therefore, be placed upon the Mineral Map. The famous furnaces (1" 177, N.W. corner) were erected where the *n* of *Araglin R.* is engraved. They are carefully marked on Griffith's Map of 1855.

Drumslig. 1" 188. 6" Waterford 35 S.E. Haematite occurs here in the joints of shaly beds of white sandstone of the Upper Devonian series (Mem. 188 and 189, p. 21, 1861). The deposit was worked by Sir Walter Raleigh, who was then resident at Youghal, in 1600 ; Stewart (p. 136) noted the old pits in dark iron ore in 1800, and mining was again carried on between 1850 and 1860, when operations ceased. See Kin. pp. 49 and 105. Mem. 188 and 189, p. 21, states that ore was also worked in **Grallagh** (6" 38 N.W. and N.E.). There are townlands of Grallagh Upper and Lower in 6" 38 N.W., and the mine may have been on the bank of the Licky River in one of these.

Rostellan. 1" 195. 6" Cork 88 N.E. A highly siliceous earth, sometimes styled " Rostellan clay," has been excavated at Rostellan, on the eastern side of Cork Harbour. In the Geological Survey Memoir on the Cork district, p. 112 (1905), an occurrence of "brown haematite" 5 to 8 feet thick is recorded in connexion with the siliceous material. This deposit seems never to have been exploited, and verification of its thickness and extent is desirable.

Coosheen. 1" 199. 6" Cork 139 S.E. and 140 S.W. This is one of the copper mines of the Skull district; but Kane (1845, p. 126) speaks of it as a possible source of haematite.

Aghadown. 1" 199. 6" Cork 140 N.E. and S.E. Kane's reference to haematite at Coosheen, mentioned above, adds interest to a statement in Smith's "History of Cork" (1750), quoted in Mem. 200 etc., p. 26. It appears that at one time iron ore was raised in the parish of Aghadown, north of Roaring Water Bay, where Ba is marked in colour on the Mineral Map. It was in sufficient quantities to support considerable furnaces. Compare with Araglin.

Aghatubrid, Glandore district. 1" 200. 6" Cork 142 N.E. and 143 N.W. The manganese mines here yield a considerable quantity of haematite. Min. Stat. 1860 record 60 tons 13 cwt. of "brown haematite" as raised from Glandore. Kinahan, in Mem. 200 etc., p. 25 (1861), says the works were discontinued when he visited them; but this may refer to manganese working only. He quotes a good account of the lode written by Wyley in 1854. The haematite is said to increase in depth, replacing the ore of manganese. The working of the manganese revived in 1876. (See under Manganese).

Bog Iron Ore Localities.

Bog Iron Ore is raised from time to time at various places in Ireland (see Kin. pp. 40 and 64), but is recorded only by the port of export. Ports such as Westport, Ballina, and Dublin, or Dublin, Wicklow, and Arklow, are grouped together in Min. Stat., and the sources of the ore are thus rendered more obscure. The record of 500 tons of ore from Buncrana in 1872 probably refers to bog-ore, and explains the mention of Buncrana as an iron-producing locality in 1866. In Min. Stat. 1866, 1,006 tons of "argillaceous carbonate" iron ore are recorded from *Ballnnass*. This locality appears again in 1868 as *Balliness* (800 tons), and in 1880, with 1,051 tons, valued at £630 12s. Mr. Lyburn points out to me that the reference is probably to the export of bog iron ore from the pier of Ballyness N.E. of Muckish Mountain in County Donegal (6" Donegal 24 N.E.). The 6" MS. maps of this district contain references to considerable deposits underlying peat. As examples of the fluctuations in the quantities exported, the following may be cited from various volumes of Min. Stat.:—1883. Dublin, Westport, and Ballina only; 8,447 tons. 1900. Dublin and

Wicklow, 395 tons; Londonderry district, including Coleraine, 1,530; Sligo with Donegal, 1,928; Westport with Ballina, County Mayo, 300. Total 4,153 tons, value £1,038. 1913. Dublin, Wicklow, and Arklow, 1,380 tons; Londonderry, 2,455 tons. No record from Westport and Ballina.

The western trade is, no doubt, supplied in part from the bog-deposits of the **Feakle** district in County Clare, 1" 124. See Mem. 124 and 125, p. 48 (1863). Feakle is to be found on the Mineral Map just below the name Kildavin.

Several of the localities of bog iron ore in COUNTY DONEGAL are given in the Memoirs to sheets 1, 2, etc., p. 36, and to 3, 4, 5 etc., p. 117. See also Ballyness above. A note on the MS. 6" sheet of COUNTY GALWAY, 65, in the Geological Survey Office states that bog ore occurred in the townlands of Gortmore, Glencoh, Turlough, and Derravonniff, and that an old iron furnace stood in Camus Oughter. One of the most famous furnaces for bog ore was at **Woodford,** 1" 125, 6" Galway 125 S.E., where smelting was carried on down to 1750. Hely Dutton (Stat. Surv. Galway, p. 33, 1824) says that some of the ore was raised locally, and C. Giesecke ("Account of a mineralogical excursion to the counties of Galway and Mayo," Proc. R. Dublin Soc., vol. 62, appendix to Proc. of 2nd March, 1826) states that it was "meadow iron ore" and dug from under turf. See also Kin. pp. 74 and 84.

CHAPTER X.

LEAD.

When it is observed that the copper ores of Ireland are mostly associated with rocks of Devonian (Old Red Sandstone) or older age, while the ores of lead and zinc occur largely in the Carboniferous Limestone, this does not imply a difference in the epoch of deposition in the lodes. As A. M. Finlayson has pointed out ("The Metallogeny of the British Isles," Quart. Journ. Geol. Soc. London, vol. 66, pp. 287 and 292, 1910), almost all the important mineral occurrences in the British Isles may be due to the infilling of fissures caused by the Armorican movements at the close of Carboniferous times, the difference in distribution of the materials depending on a zonal distribution in depth. The tin ores conspicuous in Cornwall represent the lowest zone, where the action of vapours from the deep-seated igneous cauldrons was intense; copper and gold occur above; and lead and zinc still higher. While (*ibid.*, p. 284) the pyritic deposits of the county of Wicklow seem associated with the post-Silurian but early Devonian (Caledonian) movements and intrusions, and while Messrs. Hill and Collins (quoted by

Finlayson, *ibid.*, p. 291) have urged that the lead and zinc ores of Cornwall are of Cainozoic age, Mr. Finlayson believes that both the lead and zinc, and the great bulk of the copper, are Armorican. Subsequent modifications in the way of oxidation, partial removal, and rearrangement must have occurred, and fissures connected with the Cainozoic Alpine movements may have become enriched by products from the earlier lodes.

The mining of lead has naturally been connected with that of silver, as the old-world symbols of the Moon and Saturn show so frequently on Griffith's Map. A mine of galena is thus often marked as " Silver Mine " on the old engraved 6″ sheets of the Ordnance Survey, and at times even on recent 1″ sheets, and the well-known town of Silvermines in the County of Tipperary records this valuable inducement to the mining of the duller lead. In the Ovoca district of Wicklow, residual strings and concentrations of metallic silver have been found in the upper " gossany " parts of lodes that formerly contained galena. The mine in Co. Antrim stated by Boate (" Ireland's Naturall History," p. 141, 1652) to be undeveloped but very rich, and yielding a pound of silver to every 30 lbs. of lead, has unfortunately vanished into mere tradition.

Galena, the common ore of lead, contains as much as 86·6 per cent. of the metal, and to obtain one ton of lead we need 1·1547 tons of pure ore. The silver-content of the ore is usually calculated for commercial purposes as so many troy ounces of silver to the ton of galena. Formerly it was often stated as oz. per ton of extracted lead (Kane, 1845, and Hunt in Percy, " Metallurgy of Lead," p. 102, 1870). Ten oz. per ton of galena represents a moderately argentiferous ore; this would yield 11·547 oz. per ton of lead and, there being 32,666·66 troy oz. in a ton, 0·0306 per cent. of silver in the ore. If the percentage of silver in the galena is given, multiplying by 326·7 gives the troy oz. per ton of ore. The silver is commonly regarded as being combined with sulphur as argentite, Ag_2S, which is isomorphous with galena; but it may also exist as patches or streaks of the native metal absorbed in the galena.

The great number of occurrences of galena recorded throughout Ireland, especially in veins in the widely spread Carboniferous Limestone, has made it necessary for the purpose of the present Memoir to examine the credentials of so-called mines even more closely than in the case of copper pyrites, and as far as possible to restrict the names on the map to cases where serious workings have been undertaken. Owing to local smelting of the ore, the output quoted from Min. Stat. is here given as actual metal in cases where this has seemed to be of more interest than the totals of ore raised. The Report of the Controller of the Department of Mineral Resources for 1918 (p. 17) states that one ton of dressed ore from a mine in Co. Armagh represented all the lead raised in 1916 in Ireland. The only lead mine recorded as working in the Home Office List of Mines for 1918 is that at Abbeytown in Co. Sligo.

Glentogher (Carrowmore). 1" 5. 6" Donegal 20 S.E. The engraved 6" sheet marks "Silver Mines" slightly to the west of the road leading from Carndonagh to Carrowkeel upon Lough Foyle. Mem. 1, 2, 5, etc., p. 36 (1890), mentions old shafts sunk here, about which no information could be obtained. C. Giesecke (Appendix to Proc. R. Dublin Soc., 14 Dec., 1826, p. 13) noted 6 old shafts, said to be connected by a level, and worked about 1780. Griffith (1861, p. 143) says that zinc ore was present. The mine was revived from 1905-6, raising 400 tons of ore in the former year (95 tons lead, with 4,000 oz. of silver, that is, 42 oz. to the ton), and 1,400 tons of poorer ore in 1906. The total value was £7,600. The proprietors were Messrs. R. and J. Johnson of Belfast.

Keeldrum (*Kildrum*). 1" 9. 6" Donegal 33 N.E. This mine occurs in the complex country of metamorphosed pre-Cambrian sediments and granite intrusions that constitutes the highlands of Donegal. It lies close to the high road running southward from the north coast to Gweedore, and is two miles south of Bedlam; a lode and symbol are marked on the 1" Geological Survey Map. Other lodes are indicated to the north. Mem. 3, 4 etc., p. 115 (1891), describes the occurrence of silver-lead ore, and the mine as "opened many years ago" and long abandoned. The Reports of the M.C.I. show that it was started early in 1826, when it was regarded as "likely to yield to few in this country." C. Giesecke (Appendix to Proc. R. Dublin Soc., 14 Dec., 1826, p. 7) describes it as then 27 fathoms deep, with a level 500 feet long. Later in the year 54 tons of ore were sent to the smelting works at Ballycorus, Co. Dublin, a considerable journey in those days. A further quantity of 120 tons of "extremely rich" ore was raised (M.C.I., 1826, third report). Later outputs were 335 tons of ore in 1827, with improving prospects, and 60 tons a month in 1828. Water-troubles then set in; a 50 feet wheel was sent for to work the pumps, on a shaft of 40 fathoms; but in 1832 work was suspended and the machinery was transferred to Luganure. In 1860, however, the Company revived "Kildrum," carried on work energetically for a year or so, but abandoned the lode in 1862. According to Min. Stat. Lists of Mines, they retained it in their hands until 1865. The mine was thus idle when Kane wrote (1845, p. 210). Holdsworth ("Geology etc. of Ireland," p. 102, 1857) adds "ochre and iron" as occurrences at Keeldrum. Kinahan (p. 15) says that the lode was worked out; but its history suggests further possibilities in a district that is now more accessible.

Keeldrum was reported on by Mr. H. J. Daly to the Department for the development of Mineral Resources in 1917.

Drumreen. 1″ 10 (extreme north, Long. 7° 47′ 10″ W.). 6″ Donegal 27 N.W. A mine of argentiferous galena existed here about the middle of the nineteenth century, though there is no indication on the Geological Survey maps. The hamlet of Drumreen is in Carrickart townland, in the mica-schist country east of Sheep Haven, and under the *k* of *Carrickart* on the Mineral Map. " Silver Mines " is engraved here on the 6″ Ordnance Survey sheet, though not on the first issue of 1836. Kinahan (p. 15) records lead ore here. Mr. H. J. Daly reported on the occurrence to the Department for the development of Mineral Resources in 1917.

This locality may have supplied the lead ore, said to be from a vein 4 feet wide, that was smelted by peat fires by the Donegal fishermen for net-sinkers (Stewart, p. 40, 1800).

Glenaboghil (Fintown). 1″ 16 (western margin). 6″ Donegal 58 S.E. The site, which is among the gneisses of a moorland district, is shown as that of a " silver mine " on the 6″ and 1″ Ordnance Survey sheets, in Glenaboghil townland. The mine was at the S.W. end of the little lake, in a hollow about a mile N. of Fintown, and is approached by a small track from the S.E., turning off the main road near Mill Bridge. The lode is shown on the 1″ geological map. Another vein of lead ore occurs not far away on the S. shore of **Loughnambraddan** (1″ 15. 6″ 66 N.E.), and the coloured spot for the name Fintown indicates this locality on the Mineral Map. Glenaboghil mine lies 1 mile to the N.E. Griffith, possibly by some confusion of the two veins, styles Glenaboghil " Fintown (Loughnambraddan)," (1861, p. 144); but the site is marked correctly north of Fintown on his Map. Glenaboghil was opened before 1826, when Sir C. Giesecke reported on it (Appendix to Proc., Royal Dublin Society, 14 Dec., 1826, p. 11) as " formerly worked by some English miners." He found the galena much mixed with blende, but regarded the mine as very promising. Mem. 3, 4 etc., p. 115 (1891) implies that this mine was worked as recently as 1850, and that difficulties of transport led to its abandonment; ores of iron and manganese are recorded here.

Castlegrove (Eighterross). 1″ 17 (extreme N.W. corner, above Castle Wray). 6″ Donegal, junction of sheets 53 N.E. and 54 N.W. Lead ore occurs here on the shore of Lough Swilly, about 4 miles N.E. of Letterkenny. The M.C.I. took up a lease in 1851; but nothing more seems to have come of it. Griffith's Map has a lead sign here, and he regarded Eighterross as a mine (1861, p. 144). Kinahan's **Knockybrin** mine (note on MS. 6″ sheet Donegal 53 N.E. and Mem. 3, 4 etc., p. 117) lies immediately to the west. His lead indications appear in the extreme S.E. corner of the geological sheet 1″ 10, and it is possible that the M.C.I. actually worked in Knockybrin.

Kilrean. 1″ 23. 6″ Donegal 74 S.W. This mine was in Kilrean Lower, on the south side of the road from Glenties to Ardara, close to the Owenea, and 1½ miles S.W. of the present hamlet of Kilrean. It is marked as a lead mine on the engraved 6″ map. It is named *Kilbrain* in Min. Stat. Lists of Mines 1860-65, with C. Clemes as proprietor in 1860, and thenceforward the M.C.I.

Mullantiboyle. 1″ 23. 6″ Donegal 74 N.E. Stewart (p. 49) notes in 1800 that lead ore was raised about 50 years before " near the inn in Glenties," the works being stopped by water. J. McParlan (Stat. Surv. Donegal, p. 25, 1802) says the working was in " Mullentybogh in Glantice " and that the " Onea " (Owenea) River broke in. Giesecke (Appendix to Proc. R. Dublin Soc., 14 Dec., 1826, p. 8) speaks of a lead mine close to the river near Glenties worked to 10 fathoms a few years before his visit, " but without good success." There may thus have been a reopening since 1750. The MS. 6″ map of the Geological Survey marks a quarry south of a bend of the Owenea, in Mullantiboyle, a townland close to and S.W. of Glenties, which may be a relic of Stewart's, McParlan's and Giesecke's lead mine. McParlan also mentions mines, probably trivial, at " Norin " (Narin), Drumnacross, and, vaguely enough, one on the " middle mountain," which may be Fintown.

Ballyshannon Mines. 1″ 31. 6″ Donegal 107 N.W. Griffith (1861, p. 143) records a number of lead mines which were at one time worked near the coast north and west of Ballyshannon, including **Abbey Island** and **Abbeylands,** and also **Ballymagrorty** (6″ 103 S.E.), 2 miles south of Ballintra. They were recognised in Min. Stat. Lists of Mines from 1860–5. Memoir 31 and 32 (1891) mentions some of them; but the sites are now concealed by farmlands. The only output recorded seems that given by Kinahan (p. 81) of iron-ore taken from the **Carricknahorna** lead lode in 1884. Barytes occurs in some cases.

Twigspark. 1″ 43. 6″ Leitrim 7 S.E. and S.W. In this townland, 2 miles N.W. of Manor Hamilton and ½ mile W. of Lurganboy, the words Silver Mine occur twice on the engraved 6″ sheet, probably representing workings on the same lode. They are repeated on the 1″ sheet published in 1877, and the 1″ geological map indicates silver, zinc, and lead. Stewart (p. 91) and McParlan (Stat. Survey Leitrim, p. 14, 1802) knew of the occurrence; the latter says the ore was quarried. The mine was working again from about 1842-6 (Mem. 42 and 43, p. 29, 1885).

The MS. 6″ map of the Geological Survey records ores of lead, copper, and zinc in Twigspark, with specks of native silver in the vein-stuff of the western mine. The ore is in dolomitic Carboniferous Limestone.

The Down Lead Mines.

Several workings for lead have been made on lodes that penetrate the Silurian shales and sandstones of the county of Down. A considerable lode runs north of the town of Newtownards, on which two mines were opened, Conlig and Newtownards.

Conlig (*Clonligg*; **Whitespots**). 1" 37. 6" Down 6 N.W. This mine was in Whitespots townland; it was idle in 1802 (Dubourdieu, Stat. Surv. Down, p. 12, 1802), after being worked some years before. Giesecke (Appendix No. 1, Proc. R. Dublin Soc., vol. 67, p. xiv., 1831) says that it did not pay; but it was revived when Kane noted it in 1844 (1844, p. 199). The first record of output is in 1845 (Hunt 1848). In 1847, 208 tons of lead were produced, but only 40 tons in 1852. Work stopped in 1853. The Geological Survey in 1865 (Mem. 37, 38, 39, p. 43, 1871) was unable to do more than gather information from the miners. The lode was again worked between 1880 and 1885 (see Newtownards below). S. Haughton ("Account of the Gangue of Conlig Lead Mine," G.S.D., vol. 5, p. 203, 1853) described the gangue of the lode as "a pure hornblende," on the basis of analyses by himself and Galbraith. G. W. Lamplugh (Mem. on the Geology of the country around Belfast, p. 123, 1904), from an examination of material on the spoil-heaps, regards this part of the lode as a Cainozoic basalt dyke, a suggestion that is of great interest in assigning a late geological date to the importation of the lead ore into the rocks of Co. Down.

Newtownards. 1" 37. 6" Down 6 N.W. This mine, which is engraved on the 6" map, was opened on the south end of the Conlig lode, and is not separately mentioned in the Geological Survey memoirs. Stewart (p. 59) speaks of mine-holes and a level made here a few years before 1800 by a Company that raised a considerable amount of ore. Distinct returns are given from Newtownards in Hunt 1848 and Min. Stat. from 1845 to 1865. In 1852 as much as 1,420 tons of lead were produced, and 1,084 tons in 1854; but in 1860 only 169 tons of ore were raised, yielding 128 tons of lead. A break occurs from 1866 to 1880. From 1880 to 1885 small quantities were raised by the Newtownards Mining Company, and a further attempt was made in 1899, when the 82 tons of lead produced were valued at £228.

Castleward. 1" 49. 6" Down 31 S.E. The best record of this mine, which was on Dickson's Island in Castleward demesne, is that by W. A. Traill, in Mem. 49, 50, and 61, p. 66 (1871). The only information elsewhere appears to be the records in Min. Stat. for 1862 (85 tons lead) and 1863 (18½ tons lead). Traill, however, says that the mine was opened in 1855, so that the entry under "Strangford" in Min.

Stat. for that year (23 tons lead) probably refers to it, though Strangford is there said to be in County Antrim. The shaft reached 30 fathoms; but the ore was too irregular to pay. Zinc blende was present. The mine is named in Min. Stat. Lists of Mines from 1860-5.

Tullyratty. 1″ 49. 6″ Down 31 S.E. There was a small mine in the townland of this name, S.W. of Castleward demesne; it is marked on the engraved 6″ map. No output is recorded; but Kane (1845, p. 218) gives the silver-content of the ore as 10 oz. to the ton of lead. Griffith records copper as well as lead (Map and 1861, p. 144). See Traill, Mem. 49, 50 and 61, p. 66 (1871).

Dundrum. 1″ 61. 6″ Down 44. C. Giesecke (Appendix No. 1, Proc. R. Dublin Soc., vol. 67, p. xiv., 1831) mentions a lead mine as worked on a galena indication near Dundrum, which did not pay.

Coney Island. 1″ 42 and 54. 6″ Sligo 7 S.E., 8 S.W., etc. Gerard Boate in 1645 ("Ireland's Naturall History," p. 141, 1652) knew of only three lead and silver mines in Ireland, one of which was " upon the very Harbour-mouth of Sligo, in a little Demy-Iland commonly called Conny-Iland." It was rather an occurrence of lead ore than a mine, since it had been discovered only a few years before 1641, and remained undeveloped; but the note seems worth recording. Coney Island, connected with the mainland on the south by sands at low water, lies across the mouth of the Sligo inlet. The MS. 6″ map of the Geological Survey, Sligo 8 S.W., shows a string of galena one inch thick at Deadman's Point on the Rosses promontory opposite Coney Island. This may have been the " mine "; but Boate's informant may have confused Coney Island with the occurrence of abundant lead ore at Abbeytown, Ballysadare.

Abbeytown (Ballysadare). 1″ 55. 6″ Sligo 20 N.W. The lode is in Carboniferous Limestone, on the southern shore of the head of Ballysadare inlet. Kinahan (p. 98) and Mem. 55, p. 30 (1885) speak of this mine as worked from time to time for a century past. Stewart (p. 109) mentions it in 1800 as " not wrought with spirit." In 1802 (McParlan, Stat. Surv. Sligo, p. 10) it was abandoned. It is named by Holdsworth (" Geology etc. of Ireland," p. 89, 1857) as a silver mine; the ore is said to contain 13 oz. of silver to the ton of lead. Hardman (" On a travertine from Ballysodare," Sci. Proc. R. Dublin Soc., vol. 3, p. 12, 1880) speaks of operations carried on in 1880. The mine is now likely to become known as a source of zinc. The MS. 6″ map of the Geological Survey has on its back a detailed section and notes of the mine by E. T. Hardman (see also Mem. 55, p. 31), showing the older workings. Galena and blende are here closely associated, and the mine was reopened in 1917 for both lead and zinc production (List of Mines, Home Office, 1917). It was reported on by Mr. H. J. Daly to the Department for the development of Mineral Resources in that year. See also under Zinc.

The Lead Mines of the Marches of Monaghan and Armagh.

As the number of names and mineral indications testify on Griffith's Map, many lodes have been worked on the upland of Silurian rocks that forms the main portion of the counties of Monaghan and Armagh and continues north-eastward into Down. The mines were mostly grouped round the village of Milltown, within the eastern border of Co. Monaghan, and several disused shafts, bearing the names of separate townlands, may occur on one long lode. The Tassan-Tonagh-Coolartragh lode is thus traceable for two miles (Mem. 58, ed. 2, p. 20). Records of output are generally wanting from this district. Several of the smaller ventures, not touched on here, are mentioned in Memoir 59, pp. 28-9 (1877). When Sir C. Coote wrote in 1804 (Stat. Surv. Armagh, p. 286), only one lead mine had been opened up, apparently at Derrynoose, and this was discontinued.

Clontibret Mines. 1″ 58. 6″ Monaghan 14 N.E. The church of Clontibret parish, which is ¼ mile S.E. of Milltown village, gives its name to a number of adjacent mines. Those nearest the church have been opened for antimony rather than lead, as noted by Stewart (pp. 101 and 102), who claims to have discovered the "course" in the stream-cut in 1774, and who says that rich lead ore was "poorly worked by a weak Company" near "Glentubert" church in 1800.

In the townland of **Tullybuck,** in the valley of a small stream, E. by S.E. of St. Mary's R.C. Church, and a little more than ½ mile due north of Milltown, a shaft was sunk, and two others were put down in the adjacent townland of **Lisglassan** (see map in Mem. 58, ed. 2, p. 21). Griffith (Map and 1861, p. 150) gives these names as those of separate lead mines, and he is followed by Kinahan (p. 22). The MS. 6″ map of the Geological Survey shows three shafts. A note says that the lode was 6 ins. to 2 ft. wide, and the southernmost shaft reached about 20 fathoms. The mine was reopened for antimonite in 1917. See under Antimony.

Cornamucklagh South. 1″ 58. 6″ Monaghan 19 S.E. Griffith (Map, 1855 and 1861, p. 150) gives an indication of lead ore in this townland, which lies two miles east of Ballybay. It must not be confused with Cornamucklagh North, also marked on Griffith's Map as a locality for lead-ore, and lying three miles to the north. A mine was sunk here some time before 1891, since a plan and section were then deposited in the Home Office (2589), showing small workings to 22 fathoms. No indication appears on the 6″ MS. map of the Geological Survey made about 1883, nor was the mine known to the writer of Mem. 58, ed. 1, 1885. The site of the pit is just south of the lane running S.E from the main road in Cornamucklagh South, before it turns south to the cashel.

Stewart (p. 102) in 1800 records an old valuable lead mine near "Ballyboy," and a "smelt mill" built by Mr. Pepper, but discontinued at his death.

Coolartragh (Bond ; *Coolartra*). 1″ 59 (west edge of sheet). 6″ Monaghan 14 N.E. and S.E. Coolartragh townland adjoins Tonagh on the N.E. About seven shafts (see engraved 6″ sheet and the MS. notes of the Geoolgical Survey on their copy) were opened on the lode that traverses the townland from S. to N. Griffith (1861, p. 150) is the authority for identifying Coolartragh with the Bond Mine, small outputs from which (*Bond and Newry*) are given by Hunt (1848, pp. 706 and 709) as follows :—1845, 21 tons of ore; 1846, 44 tons; 1847, 44 tons. The lead extracted was about 60 per cent. of the ore. The mine was not working (Min. Stat.) in 1854, but 15½ tons of lead are recorded from *Coolarten and Bond* in 1864. Mem. 59, p. 27, gives the date of opening as before 1847; hence the output of 1845 is probably the first. The Home Office possesses a plan, and a section down to 35 fathoms, showing four levels, by Walker and Peile of Whitehaven in 1892 (2686). The gangue of the lode is quartz with some calcite; blende is present.

Tonagh. 1″ 59. 6″ Monaghan 14 S.E. Tonagh townland lies S.W. of Coolartragh, and the shaft was in its eastern angle. Griffith (1861, p. 150) indicates the ore here as unworked, and is followed by Kinahan. A "pit" is marked on a map by Walker and Peile, of Whitehaven, in the Home Office series (2992), dated 1893. The sections show a sinking to 89 feet, and the accompanying notes state that the lessees were the Monaghan Mining Company, and that the mine was abandoned in 1893. The ore was only 3 inches thick. The Home Office list of abandoned mines (1912, p. 208) gives the mine as flooded.

Tassan. 1″ 59. 6″ Monaghan 14 S.E. This was probably the most important mine of the Clontibret district. The townland adjoins that of Tonagh on the south, and the lode is the same as that which passes northward into Coolartragh. The words "Lead Mine" are engraved west of Tassan Lough on the 1″ map of 1874, and Griffith's Map indicates the locality, which was one mile south of Tonagh mine. Griffith (1861, p. 150) names Jos. Backhouse as the first worker. Min. Stat. give the early outputs as :—1853, 16 tons lead; 1855, 35 tons; 1856 and 1857 (when the mine is said to be in Co. Louth), no record; 1858 to 1866, output culminating with 114 tons of lead in 1862. The mine was worked by the Castleblayney Mining Company from 1862-5; but it seems to have been closed in 1867. Mem. 59, p. 27 (1877), says there were two lodes, and that the workings on the eastern one reached 80 fathoms.

College Mine (Carryhugh). 1" 59. 6" Armagh 19 N.E. Two miles W. by S. of Keady, the College Mines were opened in the west of Carryhugh townland, 1 mile N.E. of Derrynoose. The lode is branched, running practically N. and S., and is clearly part of the same system that furnishes the Coolartragh-Tassan lode and those of Derrynoose and Annaglogh. The College lodes are said to have been 9 feet wide (Mem. 59, p. 29). The MS. 6" map of the Geological Survey (1873) gives several details; one of the shafts is said to have reached 35 fathoms. The gangue is quartz with a little calcite. Blende occurs with the galena. The mine is not marked on the engraved 6" map of 1836, and Min. Stat. give the first record in 1857. In 1858, 69 tons of ore yielded 42 tons of lead. The output declined to 3 tons of lead in 1864, when the record ceases.

Derrynoose (Drummeland ; *Derrynoos*). 1" 59. 6" Armagh 19 S.E. The lode is S.W. of Derrynoose village, and was mined in the southern part of Drummeland townland. Griffith (1861, p. 140) says that it was worked by Lord Farnham, and this identifies it with the mine near Keady mentioned by Sir C. Coote in 1804 as having been then the only one developed in the county (Stat. Surv. Armagh, p. 286). The vein was rich; but, when Coote wrote, the late Lord Farnham had found it hard to work without a trained manager, and his son was still awaiting aid. No profit had then been made. The M.C.I. noted the old shafts here in 1837 (see their Reports), and entered on the mine in 1838. It is described in Rep. 1838 ii, p. 5. The deep levels proved unproductive, and in 1842, there being no prospect of success, the machinery was transferred to Knockmahon. Kane (1845, p. 209) records the mine as abandoned, but says that it had yielded 200 tons of dressed ore annually. It certainly had some historic reputation, and Kane's reference may be to the working about the close of the eighteenth century. Mem. 59, p. 28 (1877), says that a depth of 35 fathoms was reached. The MS. 6" sheet of the Geological Survey shows the sites of three shafts, and the words "Old Mines" are engraved. Blende and iron pyrites occur with the galena.

Clay (*Clea*). 1" 59. 6" Armagh 19 S.E. (extreme eastern edge). This small mine is not marked on the Mineral Map, but lies just east of the Pb of the Derrynoose mine, and southwest of Clay Lake. It is marked as "Old Mines," with sites of two shafts, on the engraved 6" sheet, and as "Lead Mines" on the engraved 1″ sheet of 1874. The M.C.I. took up this lode in 1826 (Rep. for 1826 i), but abandoned it later in the year. Mem. 59, p. 29 (1877), is unable to provide any information as to the lode. Griffith (1861, p. 140) says that manganese ore occurs.

Annaglogh. 1″ 59. 6″ Monaghan 15 S.W. On the 1″ Geological Survey map, the lode in Annaglogh, one mile N.E. of that of Tassan, seems interestingly associated with a Cainozoic dyke, which has a more northerly trend than is usual (compare with Lamplugh's remarks on Conlig). The MS. 6″ map of the Geological Survey gives some details of the shafts, one of which was sunk as much as 40 fathoms to the lode. The only records of output from Annaglogh are in 1852 (310 tons of lead) and 1853. The mine is mentioned, however, in 1859, and appears in the county of Clare in the Lists of Mines from 1860-5. Griffith (1861, p. 150) says it was worked by John Skimming.

Lemgare. 1″ 59. 6″ Monaghan 15 S.W. In this townland, ¾ of a mile N.N.W. from the Annaglogh mine, three shafts were put down, which are marked on the MS. 6″ map of the Geological Survey. Griffith (1861, p. 150) gives Lemgare as a worked mine. Mem. 59, p. 28, regards it as on the continuation of the Annaglogh lode.

Lisdrumgormly. 1″ 59. western edge. 6″ Monaghan 15 N.W. This mine is marked on Griffith's Map, and the lodes are shown on the 1″ sheet of the Geological Survey. The 6″ MS. map marks two lodes continuing northward from those of Annaglogh, and the western of these was reached at no great depth in Lisdrumgormly, just east of a basaltic dyke that is probably connected with this lode. In the north of the townland, close against the Armagh border, "Lead Mine" is engraved on the 6″ Ordnance sheet. A lode occurs here, also with a northerly trend; it is recorded as "rich," and is 2 to 9 feet wide. Lisdrumgormly is now under exploration by the Farney Development Company.

Croaghan. 1″ 59. 6″ Monaghan 19 N.E. and trial in 15 S.W. Griffith's Map (1855) shows lead ore at Croaghan, well to the S.E. of Tassan and north of hill 569. This corresponds with a note on the MS. 6″ Geological Survey sheet 19 N.E., in Croaghan townland, "Site of shaft sunk to work lead ore; a rich lode said to strike N. and S. with an easterly hade." Mem. 59, p. 28 (1877), records a mine with a ruined enginehouse. Griffith (1861, p. 150) gave the lode as then unworked. In the extreme east of the same townland, in 6″ Monaghan 15 S.W., the MS. map of the Survey notes that lead ore is said to have been found in a small lode just west of the streamlet that forms the border with Tattyreagh North. This spot is marked with an indication for lead on the 1″ geological map, at a point where four roads meet N.E. of hill 569.

These two localities in the large townland of Croaghan are both between the Pb sign of Tassan mine and the Pb of Annaglogh on the Mineral Map.

Tattyreagh. 1″ 59. 6″ Monaghan 19 N.E. Griffith, whose accuracy as a recorder has never been surpassed, places an indication with this name on his 1855 map between Annaglogh and the Monaghan border. The townland of Tattyreagh North, however, lies south of this, and the eastern lode in Croaghan is just across its border. It seems possible that the reference to Tattyreagh, which occurs nowhere else, may refer to what we may call Croaghan East.

Aughnagurgan. 1″ 59. 6″ Armagh 24 N.W. This mine was in Aughnagurgan townland and east of Tullynawood Lake, just above the first " g " of the name Annaglogh printed in colour on the Mineral Map. Griffith places the indication at the south end of Aughnagargan Lough, where, indeed, lead ore may have been known to him, and he gives the locality as unworked in 1861 (p. 140). Stewart (p. 19) in 1800 names a lead mine at Aughnagurgan as " poorly worked many years ago." At the Tullynawood locality, the MS. 6″ map of the Geological Survey records that rich ore is said to have been struck in a shaft at 60 feet (10 fathoms). Mem. 59, p. 29, states that iron pyrites occurs in the débris. The name " Aughnagurgan, Keady," occurs in the List of Mines for 1916, the mine being said to be suspended. It was then in the hands of Mr. Robert Espinasse, of Dundalk.

Tamlat. 1″ 69. 6″ Monaghan 22 N.E. There is a lead indication on the MS. 6″ Geological Survey sheet in this townland, 100 yards east of the border of **Aghnamullen.** A mine may have existed in one of these townlands, since Stewart (p. 102) states that potter's clay and lead ore were worked with great profit in " Aghnamullan " about 1750. The locality is north of Cootehill.

The Hope Mines (Cornalough and **Carrickagarvan).** 1″ 70. 6″ Monaghan 25 N.W. These mines were in Silurian strata about two miles south of Castleblayney, west of the road from that town to Carrickmacross. The lodes will be found on the 1″ Geological Survey map. One of the shafts (Mem. 70, p. 34, 1877) reached 100 ft. (17 fathoms). Griffith (1861, p. 150) gives the minerals as argentiferous galena and barytes. This is probably the *Dundalk* mine of Min. Stat., since the Hope is quoted in 1864 (p. 38) as " Louth : Hope (Dundalk)," and the Dundalk mine is given as in Co. Armagh in the Lists of Mines, 1867-78. The outputs under Dundalk are 1852, 38¾ tons lead ; 1859, 14 tons. Under Hope we find 1864, 19 tons lead (215 oz. silver) ; 1868, 53 tons ; 1869, 17 tons. There is no later record. It is possible that the *Castleblaney* mine, placed in Co. Clare in Min. Stat. for 1849, is an early record of the Hope, with 21 tons of lead. A Castleblayney mine appears, however, in Co. Monaghan in the Lists of Mines, 1860-1, years in which Dundalk is also mentioned as a mine.

Creggan. 1″ 70. 6″ Armagh 31 N.W. This mine is shown, on a hillside ½ a mile N.E. of the village of Creggan, on the engraved 1″ sheet of 1874, and is called " Old Lead Mine " on the issue of 1902. It was unknown to Griffith in 1861, and apparently to Kinahan, though it was described in Mem. 70, p. 34, in 1877. The ore was argentiferous, in a gangue of barytes and quartz. No record of output seems available.

Lochstuckagh. 1″ 59. 6″ Cavan 17 S.E. The MS. 6″ Geological Survey map shows a lode here, with the note " worked 40 years ago," which would be about 1833. The townland is east of Cootehill and S.W. of Mayo Hill ; a gold line occurs on the 1″ map. This is probably the lead mine said by Sir C. Coote to be in Mayo townland, which lies immediately to the east (Stat. Surv. Cavan, p. 241, 1802). Coote says it was worked about fifty years before the date of his book, which would carry it back to 1750, and that it was closed because the owner wanted more than one-fourth of the ore that was to be smelted on the spot.

Salterstown. 1″ 82. 6″ Louth 16 S.W. This occurrence has been referred to under Copper. Mem. 81, 82, p. 33 (1871), says that the shaft in search of lead ore was sunk near the old church in 1830 by the Hibernian Mining Company, south of their copper trial, but was soon abandoned.

Sheeffry (*Sheffry, Sheffrey* ; **Tawnycrower**). 1″ 84. 6″ Mayo 107 S.W. Very little is recorded about this mine of argentiferous lead ore, which lies in the wild country of micaschist west of the road from Westport to Leenane. It was opened on the southern slope of the Glenlaur valley, one mile north of Tawnyard Lough. William Bald, County Surveyor of Mayo, inserted " Leadmines " at this point on the map prepared by him between 1809 and 1816. The mine was lost sight of when the Ordnance Survey 6″ sheet 107 was first issued in 1839 ; but it appears later as " Silver and Lead Mines " on the 1″ sheet. Griffith (Map, 1855 and 1861, p. 149) knew it as Tawnycrower, from the name of its townland, and as a worked mine. C. Giesecke visited it in 1825 (Appendix to Proc. R. Dublin Soc., 2 March, 1826), when the Hibernian Mining Co. was at work here. The level was then driven to 27 fathoms. Some copper pyrites occurred with the galena. Holdsworth (" Geology etc. of Ireland," p. 88, 1857) looked forward to its being reopened by the newly founded West of Ireland Mining Co. He says that the percentage of silver was high, and that the works had been long deserted. The mine had been developed by a series of adit levels driven from the slope of the hill. The lode ran nearly north and south. There is, curiously enough, no mention of the mine in Mem. 84. The Home Office possesses a plan in the abandoned mine series (R. 293).

Derrylea. 1″ 93. 6″ Galway 36 N.W. This mine was east of Derrylea Lough, 3½ miles east of Clifden, S. of the highway to Oughterard. It was in the metamorphic rocks, and was, according to Griffith (Map 1855, and 1861, p. 145), worked for lead by Messrs. Gibbs, Baxter and Reynolds. Kinahan, however, perhaps accidentally, indicates it as unworked (p. 18), and mentions an unrewarded search for gold, probably in connexion with the iron pyrites that occurs (see under Sulphur). No record of output seems to be available. The mine is not marked on the 1″ geological map, and has not been inserted in the Mineral Map. Its site on this map is at the " o " of CONNEMARA. It is mentioned here chiefly to avoid confusion with the Derrynea mine, some 12 miles to the S.E. It was examined and freed from water in 1918.

The Clements Mine (Carrowgarriff). 1″ 94. 6″ Galway 39 N.W. This locality, 2 miles S.E. of Maum, and ¼ mile N. of the most westerly prolongation of Lough Corrib, is in a region of mica-schists; the lode was unknown to Griffith, and its development seems to have been quite modern. Kinahan (p. 19) gives Carrowgarriff as unworked (1889). The lode is marked on the 1″ Geological Survey map. Mem. 93, 94, p. 164, mentions a trial here. Min. Stat., under the name Clements, *Carrowgarow*, quote 2 tons of ore valued at £12, as raised in 1908. Stewart (p. 79) mentions a lead mine working in 1800 " in the face of the mountain " at the head of Lough Corrib. This must have been near the Clements lode.

THE LEAD MINES NEAR OUGHTERARD.

A number of lodes occur here, mined under various names, both in the metamorphic and granitic series and in the Carboniferous rocks. It may be presumed that all the orebodies are later in date than Carboniferous times. The outputs seem to have been small, and, by the lack of reference in Kane, development took place mostly after 1844. Hely Dutton (Stat. Surv. Galway, p. 30, 1824) mentions an opencast, and the raising of 3 tons of lead ore on Mr. French's land a few miles from Oughterard barrack.

Glengowla (*Glengola*). 1″ 95 (near S.W. corner). 6″ Galway 54 S.W. This lode is south of the high road to Oughterard from Clifden, in the townland of Glengowla East and in metamorphic rocks. The gangue is calcite and barytes; blende and fluorspar occur. The mine is described, with a plan and section, in Mem. 95, p. 65 (1870). Min. Stat. give it as worked in 1851 and 1852. Thirty-nine and one-quarter tons of lead from it yielded 140 oz. of silver, or 3·57 oz. per ton. The Lists of Mines give it as a mine of lead ore and pyrites under Mr. G. F. O'Flahertie (O'Fflahertie) from 1860 to 1865. It was reported on by Mr. H. J. Daly to the Department for the development of Mineral Resources in 1917. - See also under Zinc.

Claremount. 1" 95. 6" Galway 54 N.W. This mine, east of Glengowla and close to Oughterard, is marked as a copper mine on the engraved 6" map, and has been already mentioned under Copper. It was in the south-east of Claremount townland, and just north of the high road from Clifden to Oughterard. The shaft sunk by Colonel Martin (Mem. 95, p. 64, 1870) was at this point; its date is not recorded. Claremount is marked by Griffith as a lead locality on his Map (1855), and as a worked mine (1861, p. 146).

Canrawer West. 1" 95. 6" Galway 54 S.W. Griffith marked this mine in 1855, and it is that described, including " O'Fflahertie's shaft," in Mem. 95, p. 64. It was south of the high road and the river, in the western part of Canrawer West, between the coloured spots indicating Claremount and Cloosh on the Mineral Map.

Oughterard. 1" 95. 6" Galway 54 N.E. This appears in the Lists of Mines, 1872-3, as worked by the Galway Mining Company, and farther back in Min. Stat. 1860, as the name of a mine raising 11½ tons of ore. The spot marked with this name, just east of the town of Oughterard, in the Mineral Map is the site of the **Lemonfield** shaft, which was in the west of the townland of that name. This was worked by G. F. O'Fflahertie; the ore was argentiferous (Griffith, 1861, p. 146, and Map 1855). Mem. 95, p. 63, says that the shaft was only 3 fathoms deep, but the ore improved as it was followed down. Stewart (p. 80) mentions a small pit and opencast with rich lead ore at or near Lemonfield in 1800.

Cregg. 1" 95. 6" Galway 54 S.W. Griffith (1861, p. 146 and Map) recognised a lead mine in this townland, south of Canrawer West. Mem. 105 and 114, p. 58, gives it as worked in 1865-6, and rich in copper pyrites. The lode and two shafts are shown on the MS. 6" Geological Survey map, in the extreme S.W. of the townland.

Cloosh (Clooshgereen). 1" 105 (extreme north). 6" Galway 54 S.W. The townland of Clooshgereen is in the gneissic region south of Oughterard, and the mine was on the continuation of the Cregg lode. A plan and section of it are given in Mem. 105 and 114, p. 59 (1869); it seems to have been worked about 1861, and was not known to Griffith when he drew up his "Catalogue of Mines." The gangue was largely barytes (see under Barytes). The Lists of Mines name it (as *Cloost*) in 1876 and 1877. A MS. plan of Cloosh, dated 1866, is in the office of the Geological Survey in Dublin. The mine was reported on by Mr. H. J. Daly to the Department for the development of Mineral Resources in 1917.

Baronstown. 1″ 101. 6″ Meath 32 S.W. This townland lies south of Lismullen Park and east of Tara Hill. There is a quarry in it, in which no indication of ore is given on the MS. 6″ Geological map; but this may be the locality mentioned by Stewart (p. 106), where he saw in 1800 a course 5 feet wide and full of rubbish, excavated in old time for lead.

Derrynea (Cashla Bay). 1″ 104. 6″ Galway 79 S.W. The mine is west of the Cashla River, near its source in the forked lake of Glenicmurrin, in a country of bog and granite moorland. As to its possible existence in 1824, see under Rinville below. It was working when Holdsworth wrote in 1857 ("Geology etc. of Ireland," p. 85); but Mem. 104, 113, p. 86. (1871) says that the lode, with galena and iron and copper pyrites, was only 9 to 12 inches wide.

Rinville (Rinville West; Oranmore; *Renville*). 1″ 106 (S.W. angle). 6″ Galway 94 N.E. This must not be confused with the workings round Rinvyle in 6″ Galway 9 (Griffith, 1861, p. 146). Rinville lies on the east side of the bay opposite Galway town, and the mine was at the water's edge in Carboniferous Limestone. It is possible that this or Derrynea was the mine mentioned by Dutton (Stat. Surv. Galway, p. 30) as working under Messrs. Chamber and Company in 1824 on the shore of Galway Bay. Griffith (1861, p. 146) gives zinc ore and iron pyrites as associated with the lead. Holdsworth ("Geology etc. of Ireland," p. 86) writes of it as a working mine in 1857; but Mem. 96, 97, 106 and 107, p. 38 (1867), says that it was abandoned in 1849. It is named in the Lists of Mines from 1862-5.

NOTE.—Under the general title "*Galway Mines,*" Min. Stat. record 1 ton 19 cwt. of lead produced in 1852, and 3 tons in 1859. This name appears in the Lists of Mines, 1860-5.

THE LEAD MINES OF THE DUBLIN DISTRICT.

Probably on account of the proximity of a populous and enterprising city, the number of ventures in pursuit of ore round Dublin is altogether out of proportion to the importance of the lodes. The one outstanding instance of a fair body of ore was at Ballycorus, and even this owed much of its celebrity to the smelting works which were established in connexion with it, and which dealt for nearly a century with the produce of the Wicklow mines.

Cloghran. 1″ 102.   6″ Dublin 14 N.E.   J. Rutty (" Natural History of the County of Dublin," vol. 2, p. 138, 1772) records " very rich ore " as raised from two mines at the church of Cloghran, on the road from Dublin to Swords. " Lead Mine " appears on the engraved 6″ Ordnance Survey sheet in the eastern part of the large quarry in the limestone. Griffith's Map marks the locality.

Wheatfield (Church Mine). 1″ 111.   6″ Kildare 15 N.W. Griffith (" Mines of Leinster," p. 27, 1828) records a lead mine at Wheatfield, 10 miles from Dublin, where the Royal Irish Mining Company had erected considerable works. A tracing showing the mine and works, dated 1831, is in the office of the Geological Survey, Dublin. Kane (1845, p. 211) says that there was a large deposit of pure galena, which became exhausted. J. B. Jukes refers casually to the occurrence in Mem. 102 and 112, p. 71 (1875), and it was evidently not known to him except by tradition. Mem. 111 (1860) has no mention. The mine was in Wheatfield Upper, on the S.E. bank of the Grand Canal, 3 miles S.S.W. of Celbridge, and is marked in a flooded quarry in the limestone on the engraved 6″ Ordnance Survey map.

Castleknock. 1″ 112.   6″ Dublin 17 N.E.   J. Rutty (" Natural History of the County of Dublin," vol. 2, p. 138, 1772) says that Edward Ford opened a lead mine N.E. of the old castle of Castleknock in 1744. On p. 55, Rutty says that a dark ochre, suitable for paint, was found in this mine. The site is marked on Griffith's Map, and as " Old Lead Mine " on the first issue of the 6″ Ordnance Survey sheet, where quarries lie W. of the road to Castleknock.

Kilmainham. 1″ 112.   6″ Dublin 18 S.W.   J. Rutty (" Nat. Hist., Dublin," vol. 2, p. 137, 1772) says that, in eighteen months in 1767-8, 60 or 70 tons of lead ore were raised at Fleming's Quarry, Kilmainham. The lead produced was 12 cwt. per ton of ore, with 24 oz. silver per ton of lead. Working was carried down to 15 fathoms, but ceased owing to the difficulty of keeping out water. Ten tons of ore were raised from a sinking on Kilmainham Commons, which reached some 5 fathoms, and was similarly stopped by water. It would seem, therefore, that there is still ore to be extracted here ; but the Commons, with their race-course, have long since disappeared. Griffith's Map indicates the lead ore at Kilmainham, though he does not quote the locality as that of a mine (1861, p. 145). An occurrence of lead ore near at hand at Dolphin's Barn (Rutty, *op. cit.*, p. 136) caused smelting works to be set up.

Clontarf. 1″ 112.   6″ Dublin 19 N.W.   J. Rutty (*op. cit*, p. 139) says that in 1768 a mine was opened at Clontarf, on the north coast of Dublin Bay, about 80 yards from the shore, from which 14 tons of ore were raised, yielding little silver and 12 cwt. of lead per ton. The pits were only 5 fathoms deep and were filled at every high tide. Griffith ("Mines of Leinster," p. 24, 1828; also quoted in Mem. 112, p. 129, 1903) gives details of a 2 ft. vein, in which galena was associated with blende. He says that this was worked and abandoned several times since 1809. The shaft of 1809, which seems to have been a new one, reached 8 fathoms, and is still to be seen near Castle Avenue. Kane (1845, p. 210) hints that the tidal water was the difficulty, as in previous ventures, and this is confirmed by Mr. J. M. Coghlan Briscoe, a descendant of the Captain Coghlan who built the pumping station on the shore early in the nineteenth century (see his letter in the "Evening Telegraph," Dublin, 24th September, 1908, and an article in the "Freeman's Journal," 23rd September, 1908, consequent on the cutting of the old workings by the new main drainage excavations). Blasting initiated by Mr. Briscoe in 1908 showed that the lead ore was by no means worked out.

Dalkey (Mount Mapas ; Killiney Hill). 1" 112. 6" Dublin 23 S.E. There is an adit running from the sea-shore south of Sorrento Point into the granite mass of Killiney Hill, with a shaft and other traces of mining a little to the south. These represent a mine first worked in the latter half of the eighteenth century, the ore occurring, like that of Ballycorus, near the junction of the igneous mass with the Ordovician mica-schists. J. Rutty (*op. cit.*, p. 140, 1772) says that the work in *Killeny* Bay, on the estate of John Malpas, was begun in 1751. He implies that there was also a mine nearer to or at Dalkey itself. In 1818, however, Weaver (1819, p. 198) wrote of "the lead vein of Dalkey," clearly showing by his description that he was dealing with Killiney Hill. The mine was then filled by sea-water. W. Fitton, editing W. Stephens's "Notes on the Mineralogy of Dublin," p. 14, 1812), says that in 1805 it "had not been worked for several years." It was reopened about 1825 by the Royal Irish Mining Company, who worked two veins (Griffith, "Mines of Leinster," p. 17, 1828) for a brief period. Griffith marks the site as Mount Mapas on his Map, and gives Killiney Hill as an alternative name in his Catalogue of Mines (1861, p. 145). Weaver (1819, p. 198) noted the occurrence of blende and barytes, and Robert Mallet (letter in G.S.D., vol. 10, p. 70, 1863) says that the gangue consisted almost entirely of barytes. He notes also a small quantity of "antimonial lead." He says that the shafts were sunk when he was a boy, and hence the work of 1751 may have been done entirely by the adit. Though this is now flooded, barytes may still be traced in it.

The Dalkey veins, and those of Ballycorus about to be described, are probably of the same age as those of the Wicklow hills, that is, Caledonian, and therefore may be very much older than those in the Carboniferous Limestone area.

Ballycorus. 1″ 121. 6″ Dublin 26 S.W. This mine is represented by a large excavation, which was entered along the groove formerly occupied by the lode, on the north face of the hill that culminates in Carrickgollogan. The lode is in granite, and seems to split up and become lost when it enters the mica-schist that lies on the south side of the contact. It is in the east of Ballycorus townland, and above it rises the chimney built to carry off the fumes from the smelting works that lie ⅔ of a mile to the north and 400 feet nearer sea-level.

This occurrence of ore was not known to Rutty in 1772. W. Fitton ("Notes on the Mineralogy of part of the vicinity of Dublin," p. 18, 1812) describes a visit with Stephens to Ballycorus in 1807, when the workings went down for 6 or 8 feet only, were 40 yards in length, and, as now, open to the day. He says the lode ran W.N.W. and E.S.E., and was less than 2 feet wide. The ore was "small grained galena intimately mixed with grey ore of antimony." He noted baroselenite, but not the barytes that was recorded by Weaver and Griffith, and which formed an important part of the gangue. Weaver (1819, p. 198) describes two veins, coinciding above and diverging as they descend, with granite between them. The centre of the large excavation was, therefore, possibly occupied in part by granite, and not by a continuous *bonanza* of rich ore such as Smyth ("Mines of Wicklow," p. 365) suggests. Weaver found the workings extending 110 fathoms into the hill and (p. 199) 22 fathoms in depth, the veins becoming here about 20 feet apart. The mica-schist had been penetrated for 10 fathoms. The northern vein varied from a few inches to 2 feet in width, and reached in places 4 or 5 feet. The south vein was less productive, and was, in general, a few inches to 2 feet wide. The fine-grained galena from it yielded 75 per cent. of lead, and blende and pyrites occurred. It is said that traces of gold were found in the galena, and on this ground J. O'Reilly ("Notes on some assays for gold of rocks in the neighbourhood of Dublin," Proc. R. Dublin Soc., vol. 6, p. 456, 1890) had the mica-schist assayed near its junction with the granite, and in this also a trace of gold was proved. The antimonial ore mentioned by Stephens was probably local and unimportant, since it does not seem to have been traced since his time. A few stains of green carbonate indicate the presence of copper at the present day, and evidence of brown blende is frequent.

Griffith, who knew the mine by the ambitious name of *Mount Peru* ("Mines of Leinster," p. 11, 1828 and 1861, p. 145), describes the two veins, and says that the southern was still unexplored.

The following history may be gathered from the Reports of the Mining Company of Ireland for the period 1826-1863, and its publication is justified by the attraction that Ballycorus, partly through its proximity to Dublin, has offered to prospectors in later times. The Company took up the mine in 1826, when the ore was held to be not quite up to expectation. The sinking was, however, carried to 25 fathoms, probably on the shaft that is still traceable in the wood, and which is marked on the engraved 6″ map S.E. of the Shot Manufactory. The low price of lead stopped work in 1827, but smelting and rolling machinery were set up in the valley in 1828, and the first tower for the making of shot was built in 1829. Probably ore was continuously brought from the Company's mines in Co. Wicklow. In 1834 a rise in the price of lead encourgaed a new inspection of the Ballycorus venture, and some ore was raised; but in 1835 the water was allowed to run in. In 1839, after new trials, an adit was driven into the old works, and a second one in 1840; but the vein was found to be splitting up into small diverging branches, and work was again suspended. A discovery of native silver caused a new exploration in 1843, and the silver vein, very limited in its yield, was traced to a length of 6 fathoms and 8 fathoms in depth. In 1845 a cross-cut adit reached a vein like that of the old mine. It will be noted that the two veins seen by Weaver and Griffith in the old mine either merged again into one, or diverged so much that one of them might be struck in depth as a new vein. In 1845 this fresh working was stopped in unfavourable ground. The M.C.I. obtained a new lease in 1853, and actually worked the mine again in 1857, putting up a steam-engine and stamps. In 1860 further searches were abandoned. The Wicklow ore kept the smelting works in operation, and the new flue from them, extending as a tunnel built up the hill and terminating in the well-known chimney and view-tower, is referred to in the Report for 1863. The value of lead " dust " collected from this flue was £1,500 in 1865, and later estimates give £700 for each half-year. About 1863 the Company seems to have withdrawn from all speculative operations and searches throughout Ireland. W. W. Smyth ("Mines of Wicklow," Rec. School of Mines, vol. 1, pt. 3, p. 365, 1853) gives an excellent section of the works in 1853, which practically represents their final state. As previously mentioned, he attributes the great " open cast " on the hillside to a very productive bunch of ore. Presumably, this excavation was made before 1826. It was roughly filled in with granite blocks about 1919.

In the Home Office abandoned mines series (R 284) there is a neat plan of the adits and shafts of Ballycorus, made by Smyth in 1846 from a map prepared by Brockenshar for the M.C.I. It shows the site of the silver vein in the granite area, as struck in a shaft 10 fathoms below the deep adit.

The reference to Ballycorus in Mem. 121 and 130, p. 45, deals mainly with the smelting works.

Shankill (*Shankhill*). 1″ 121. 6″ Dublin 26 S.E. Quartz-veins occur as features of the junction of schist and granite on the rounded hill in Shankill townland, east of Ballycorus. Griffith ("Mines of Leinster," p. 17, 1828) says that several shafts were put down here "within the last five years," but that "no ore worth following was met with." These trials have become recorded as mines.

The Lead Mines of the Wicklow Hills.

Little, except mere detail, can be added from the point of view of the geologist or the miner to the admirable description given by W. Warington Smyth in 1853 of the lead-mining district in the County of Wicklow ("On the Mines of Wicklow and Wexford," Rec. School of Mines, vol. 1, pt. 3, pp. 349-365). The veins, as Smyth pointed out, lie near the south-east margin of the great intrusive mass of the Leinster granite, but disappear in the flanking mica-schist, which has resulted from the metamorphic action of the granite on the Ordovician shales. They seem thus to be intimately connected with the final stages of the consolidation of the granite, while the mineral belt of Ovoca, with its pyritous ores, is connected with the minor intrusions, often of a doleritic nature, that accompanied the uplift and folding of the great arch of sedimentary rocks. Both series of mineral deposits are thus probably of Caledonian age, though A. M. Finlayson ("The Metallogeny of the British Isles," Quart. Journ. Geol. Soc. London, vol. 66, p. 281) does not mention the lead ores as anterior to those connected with the Armorican movements in Ireland. The discussion by the same author of the influence of the country-rock on ore-deposition ("Lead and zinc veins in Great Britian," loc. cit., p. 321) goes some way to explain the disappearance of the veins in the mica-schist "tactite" * and in the shales beyond, considering the fissility and numerous surfaces of parting in these rocks.

The Wicklow lead mines are included in 1" sheets 129 and 130, almost all lying in 130. Smyth (op. cit., plate 1) gives a geological and mining map of the area from Glendasan to Glenmalur on the scale of 1" to 1 mile; the lodes are marked, but the names of the separate mines are not inserted. For the mines in Glendasan and Glendalough, S. Haughton ("Notes on Irish Mines," G.S.D., vol. 6, p. 168, 1855) has furnished an excellent map with names. For Glenmalur, Griffith's Map, despite its small scale, is very helpful, especially since it covers the whole length of the glen. The MS. 6" maps of the Geological Survey of course convey the most detailed information, and Mem. 121 and 130, p. 43 (1869)

---

* This useful term is due to F. L. Hess (Amer. Journ. Sci., vol. 48, p. 377, 1919), who used it for the altered calcareous rocks along a zone of igneous contact. It seems worthy of a general extension.

supplements Smyth's description of the northern mines by information furnished by J. P. Clemes, director of the works about 1867. Griffith's Map marks the whole series of mines in Glendasan merely as "Brockagh," from the upland north of the glen, and the names "Luganure" and "Glendalough" have been used by him and others both for groups of mines and for individual workings. The frequent union of the various mines under one Company has prevented a record of separate outputs, and the figures are usually given as from "Luganure" (the northern group) and "Glenmalur."

Three streams descending from the high water-parting of the Leinster range have carved romantic valleys across the junction of the granite and the Ordovician schists. The walls of these glens are set with cliffs, which are more pronounced in the schistose and shale area, where the rock-structure lends itself to the formation of ravines. The valleys widen out, as we follow them upward, into the broad basin-like forms typical of the granite moors.

The more northern valley is Glendasan (Glenasane, Glendasane, Glendassan, of older writers). The **Luganure Mines (Brockagh Mines)** apparently began here with a "small trial on lead ore" known to Stewart (p. 122) in 1800. They have been opened in the upper part of Glendasan on a series of lodes running for the most part N. and S. across the south-easterly trend of the valley (see Haughton's Map, *op. cit.*, and sheet 130 of the Geological Survey). Descending the valley from N.W. to S.E., we have the following mines :—

Luganure proper (**Old Luganure**) with **West Luganure** north of it, 6" Wicklow 17 S.W.; **Ruplagh** (*Rupla* 6" 17 S.W.; **Hawk Rock** and **North Ruplagh** were mines on the Ruplagh lode (see Smyth, p. 354); **Old Hero**, 6" 23 N.W.; **Moll Doyle**, 6" 23 N.W., and **Foxrock**, 6" 23 N.W. The distance from end to end of the series is 1¼ miles.

About a mile south of Glendasan, the Glenealo River has excavated the vale of Glendalough, which is steeply walled from its head down to the alluvial area at the foot of the Upper Lake. The stream unites with that of Glendasan at the old monastery of the Seven Churches, forming here the Avonmore. The Luganure lodes are continued through the intervening upland of Camaderry (*Comaderry*), a ridge steep to traverse; and the **Glendalough** mine proper (6" Wicklow 23 N.W.) was opened on the southern side of the ridge, where the miners' cottages still stand beyond the Upper Lake. West of it, and on the south side of the glen, the **Van Diemen's Land** (*Van Dieman's*) mine (6" 23 N.W.), so-called from its remoteness on the upland, was opened by an adit which is still traceable on the steep slope above the river, and by a shaft sunk 10 fathoms (Smyth, *op. cit.*, p. 359). Holdsworth ("Geol. etc. of Ireland," p. 23, 1857) mentions copper pyrites here.

Two miles still farther to the south, the Avonbeg, running south-east to join the Avonmore above Ovoca, has carved out Glenmalur (often written Glenmalure). Griffith (1861, p. 154), names six mines as opened here, all lying in the cliff-walled narrows of the valley above the point where the road comes over from Laragh at Drumgoff. The important work, however, was done at **Ballinafunshoge,** which is included in Smyth's map above cited, and which will be dealt with later.

Taking the mines once more in order, the Home Office series contains plans of the following as abandoned mines :— Luganure, an old plan of 1814 (R 286), including Glendasan generally, and also two plans deposited in 1887, one being of Luganure North; Ruplagh, by W. W. Smyth after Brockenshar, 1841, and also a plan of 1889; plans and sections of Hero North, Hollyrock, and Foxrock, scale 1" to 10 fathoms, 1887. **Hollyrock** is a name for a branch of the Hero lode, and occurs also in the M.C.I. Report for 1874. Ballinafunshogue, plan and section (R 284) by T. Weaver, 1812.

The history of the northern group of mines is a fairly long one. W. Fitton, editing Stephens's " Notes on the Mineralogy of ... Dublin " in 1812, says that since 1807 (p. 9, footnote) " veins of lead-ore of considerable value have been opened in the vicinity of *Glendalough*." Stewart (p. 122) in 1800 speaks only of a " small trial " as made in " Lugganure." Weaver (1819, p. 200) says that he " opened a mine in Luganure vein," and (p. 202) that he had made trials at Glendalough. We may take it that at some date between 1807 and 1812 the first step was made in the development of this important area. Glenmalur, as we shall see, was already at work. Griffith (" Mines of Leinster," p. 13, 1828) knew the " Glendelough Mines " as worked by the Mining Company of Ireland.

The figures given below are derived from the M.C.I. Reports, and also, after 1844, from Hunt 1848 and Min. Stat. The M.C.I. took up the Luganure and Glendalough property in 1826, when they raised 120 tons of ore and discovered some new veins on Camaderry. The Hero Mine was opened in 1827, was sunk to 50 fathoms, and was firmly established in 1828. **North Hero** or **New Hero** was a later venture, and, according to a note on the Home Office plan, was unconnected with the Old Hero (see under 1873 below). Hero gave 500 tons of ore in its first year, and the report of its quality in 1829 was very good.

Haughton (*op. cit.*, G.S.D., vol. 6, p. 171) gives the produce of the Luganure Mines from 1834 to 1853, from a table by Purdy Allen, Secretary of the M.C.I. From this we may judge something of the cost of working; for instance, in 1836, Purdy gives the value of ore raised as £11,506; the Report states that the profit for that year was £2,904. The mines were unproductive in 1844, on an output-value of only

£3,401; the price of lead had fallen badly. Ruplagh was now down to 110 fathoms, and was here said to be poor; workings were carried on in the upper levels only. A note by Smyth on the Home Office plan of Ruplagh of 1841 says "worked with success by the Mining Company of Ireland, but drowned in the summer of 1844, when the supply of water to work the pumping machinery failed." This may explain, more candidly than the Report given to the shareholders, why the lower levels were regarded as unproductive.

Smyth (p. 358) gives the monthly output about 1852 as 120 tons of ore, producing 74 to 75 per cent. of lead, and 6 to 8 oz. of silver to the ton. This would mean an annual output of some 1,140 tons of lead, which was by no means realised, according to Min. Stat., until 1858 and 1859, when 1,220 and 1,265 tons of lead respectively were extracted from the Luganure ores. Smelting, as usual, was carried on at Ballycorus. Thenceforward until 1865 about 1,100 tons of lead were procured annually.

About 1852 (Smyth, p. 359), Capt. Clemes discovered the N. and S. lode among the granite precipices at the head of Glendalough, where the old Luganure vein might be expected to emerge. This must have ultimately led to the opening of the **Glendalough** mine, which is possibly the "Seven Churches or Camaderry" mine of Griffith 1861, p. 153. Griffith's grouping of the mines is, however, unusually obscure. The Van Dieman's Land lode farther west must have been worked somewhat earlier, since, as previously remarked, a mine was known on it when Smyth wrote in 1853. From 1857 onward, however, the interest of the M.C.I. in the Glendalough portion of their estate was diverted from mining to forestry and cattle grazing.

In 1865 an ore styled "silver-ore" was worked from one of the veins in Glendasan, 184 tons of lead yielding 65,951 oz. of silver. This would mean a galena with as much as 0·95 per cent. of silver. The maximum recorded output of lead (1,665 tons) was reached in 1867. Min. Stat. for this year state that 1,694 tons of ore yielded 1,165 tons 16 cwt. of lead, that is, close on 69 per cent.

In 1868-70 new trials were pushed forward, improvements were made in pumping gear at Ruplagh and in the waterpower derived from the glacial tarn of Lough Nahanagan, and a new inclined railroad was made from the Van Dieman's Land mine. In 1872, however, the mines were working at a loss. Profits accrued again in 1873. A new mine was opened at Ruplagh, and the New Hero lode was proved to 50 fathoms. In 1874 a shaft was sunk near the junction of the Hero and **Hollyrock** lodes, the latter receiving mention for the first time in the Reports. In 1876 the profit on the year was over £6,000.

Lead dropped abruptly to £10 per ton in 1879; in 1881

the establishment was reduced, and Capt Kitto of the Foxdale Mines was called in to report on the conditions. By 1884, only 450 tons of ore were being raised annually, and only one lode was now being worked. In 1889 the M.C.I. offered Luganure and Glendalough for sale. In 1891 the mines were taken over by Messrs. W. H. and A. A. Wynne, who also worked Glenmalur. The output sank, however, to 6 tons of lead in 1900, after which the record ceases. New explorations were undertaken at Luganure in 1917, under the advice of Mr. H. J. Daly ("Report of the Controller of the Depart. for the development of Min. Resources," p. 40, 1918), who furnished a report under the head of Glendalough.

The mines in **Glenmalur** (*Glenmalure*) existed before those in Glendasan. Stewart (p. 122) notes "*Glenmullar*" as worked in 1800. Fraser (Stat. Surv. Wicklow, p. 18, 1801) says that this was the only lead mine then in the county of Wicklow. Stephens ("Notes on Min. of . . . Dublin," pp. 9 and 27, 1812) knew the Glenmalur mines at work in 1807. Fitton, in a footnote to Stephens, gives an output for 1811 of 334 tons of lead, when the metal was at £30 a ton. The important mine was **Ballinafunshoge** (*Ballinafinchogue, Ballynafinshogue*), 1" 130, 6" Wicklow 23 S.W. This was opened about 1800. A plan by T. Weaver is in the Home Office series, with a section showing the deep level at 80 fathoms. A large amount of the area covered had been already cleared out by 1812, when this plan was made. Weaver (1819, p. 150) shows how the granite, as often happens along its margin, had intruded in parallel sheets into the mica-schist, a feature that is well shown on the MS. 6" Geological Survey map. The lode cuts the strike of the interlaminated layers in the adit at 65°; the map shows that the deviation in general is only 40°. The course of the lode is W. 27° N., or more nearly E. and W. than Weaver estimated. The adit, which also ventilated the mine, reached the lode 85 fathoms below its outcrop on the surface. Weaver (p. 204) describes the lode as from 2 to 3 fathoms wide, the principal gangue-material being quartz, as is usual in the galena lodes of Co. Wicklow. The shaft and various buildings are named on the engraved 6" Ordnance Survey map. The annual output in 1819 was 300 to 400 tons of ore, averaging 68 per cent. of lead. Smelting was carried on in a simple manner on the spot, most of the fuel being peat. No other mine was then worked in Glenmalur. Griffith ("Mines of Leinster," p. 15, 1828) knew "Ballynafinshogue" in its early days, and says that it had been "worked by the present company for thirty years with various success." This would give 1797 as the date of opening. Stewart (p. 126) and Griffith (*op. cit.*, p. 16) note the occurrence of barytes. The ore yielded 70 per cent. of lead (see Weaver's record above). A tracing of "Dr. Lentaigne's map" of the mine is in the Office of the Geological Survey in Dublin, dated 1831. Kane (1844, p. 196; 1845, p. 208) knew the mine as working and as employing about 30 persons.

Later records are (Hunt 1848) 1845, 367 tons of ore, yielding 270 tons of lead; 1852 (Min. Stat.), 144 tons of lead and 950 oz. of silver; 1853-6, not worked; 1857-64, less than 100 tons of lead per annum. The mine was then, like others in Glenmalur, in the hands of Mr. Henry Hodgson. The record then ceases.

Smyth ("Mines of Wicklow," p. 360, 1853) describes the workings, which were then for a time idle, and mentions other occurrences of lead ore in the glen.

Clonkeen (*Clonkeene*), 6" Wicklow 23 S.W., is higher up the glen than Ballinafunshoge, on the S.W. bank of the stream. It was surrendered by the M.C.I. in 1838 (Report, 1838 ii). Griffith (1861, p. 154) records "Lead with Zinc, and Iron" here. Two lodes are shown on the MS. 6" Geological Survey map. The name occurs in the official Lists of Mines down to 1873.

Cullentragh Park (*Cullentrapark*) was a mine on the N.E. bank, opposite Clonkeen.

Baravore (*Barravore* of Griffith's Map; *Barravone* of Lists of Mines) is N.W. of Clonkeen. Smyth ("Mines of Wicklow," p. 360) speaks of a considerable trial here in 1846. The mine is in the official lists from 1865 to 1874. On the MS. 6" Geological Survey map, the lode appears to be a direct continuation of that of Ballinafunshoge.

Ballinagoneen (6" Wicklow 23 S.W., north-west corner) is on the N.E. bank, opposite and a little above Baravore. Smyth mentions working here in 1853, and Griffith (1861, p. 154) records zinc and copper in addition to lead. The mine was worked by "Sir C. P. Roney, etc." It figures as *Ballynagowen* in the List of Mines for 1871 (M.C.I. proprietors).

The Glenmalur lodes are described, from the indications on the MS. 6" maps, in Mem. 121 and 130, p. 44 (1869). All occur in a distance of about 2 miles up the glen.

Hollywood. 1" 120. 6" Wicklow 9 S.E. Griffith ("Mines of Leinster," p. 17, 1828) records "lead veins partially worked on the west boundary of the granite at the opening of Hollywood Glen." The indication on his Map lies at the north end of this picturesque cutting, south of the road from Wicklow Gap, and where the *ll* of *Hollywood* appear on the Mineral Map. Griffith reported on the lode (1861, p. 154); but it seems to have been omitted from the Geological Survey maps.

Ballycullen. 1" 130. 6" Wicklow 24 N.E. Stewart (p. 124) records a lead "mine," perhaps only a mass of ore, above the bridge of Ballycullen. The locality is vague in this large townland, which lies west of Rathnew.

Connary. 1" 130. 6" Wicklow 35 N.W. This is one of the mines of the Ovoca mineral belt, the ore occurring in the steeply dipping Ordovician slates. It has been treated of under Copper, and will also be found under Sulphur, Zinc and Antimony. One of the ore-beds of Connary was described by Weaver (1819, p. 215) as about 4 feet thick, "consisting of a fine grained intermixture of galena, grey ore of antimony, and blende, with pyrites of copper, iron and arsenic." Smyth ("Mines of Wicklow," p. 389, 1853) says that this was near the main road, and that the galena (probably he meant the mixed grey ores) averaged about 25 per cent. of lead and 30 oz. of silver to the ton. The ochreous gossan contained both silver and gold. Kin. (p. 113) speaks of this lead lode as the "Gosan lode." Outputs of lead from Connary (*Connoree*) are given by Min. Stat. in 1878, 133 tons 10 cwt. of lead, value £1,882, and in 1885, when 20 tons of lead and 980 oz. of silver were obtained from the "bluestone" by the Ovoca Mineral Co.

Knockanode. 1" 139. 6" Wicklow 35 N.W. See under Sulphur, Ovoca Mineral Belt.

Ballintemple. 1" 139. 6" Wicklow 40 N.W. This mine was in Ballintemple townland on the south side of the Aughrim (Darragh or Darach) River, one mile above Woodenbridge, where the *R* of *Aughrim R.* is on the Mineral Map. The lode runs some 30° N. of W., and contained a vein of continuous galena 1 to 8 ins. thick (Smyth, "Mines of Wicklow," p. 397, 1853). Kinahan (p. 481) disputes the occurrence of an 18-inch vein mentioned in Mem. 138 and 139, p. 31. The workings were never extensive. See also Carysfort below. Ballintemple must not be confused with the townland of the same name near Newtown Hamilton in Co. Armagh, which is also a locality for lead ore.

Carysfort. Mines in southern Wicklow on Lord Carysfort's property were often styled the Carysfort Mines. The townland of Carysfort, it may be remarked, is near Kingstown (Dunleary) in Co. Dublin. A *Carysfoot* mine appears in Min. Stat. for 1860, p. 175, and also in 1861; it is placed in Co. Waterford. In 1862-8 it is given in the Lists of Mines as Carysfort in Co. Wicklow, raising lead ore and pyrites. Outputs obtained here by the Carysfort Mining Company are given for 1860, 1861, 1864, 1865, and 1867, culminating with 31 tons of lead in 1864. It seems possible that this refers to Ballintemple; but Kinahan (p. 483) mentions a venture by the "Carysford Mining Company" apart from the occurrence of ore at Ballintemple. This, however, does not seem to have resulted in a mine.

Arklow appears as the name of a mine in Min. Stat. 1850, where it is placed in Co. Clare. In 1851 it moves to Wicklow. The output annually was only 14 tons, and it possibly represents shipments from the port.

Wicklow appears as the name of a mine in Min. Stat. 1850, with an output of 52 tons 11 cwt. of lead. Luganure is separately returned, and the source of this lead is not now traceable.

A lead mine named **Avondale**, Rathdrum (Co. Wicklow), is mentioned in Min. Stat. Lists of Mines, 1874-7. The output seems unknown.

Lead Mines of Southern Galway.

Ballymaquiff. 1″ 115. 6″ Galway 113 S.E. A lode is marked on the MS. 6″ Geological Survey sheet in Ballymaquiff North, with the note "a large mass of white calc spar with the ore in 3 small pockets." Mem. 115 and 116, p. 39 (1865), says it was worked about 1850. It appears in the Lists of Mines 1862-5, but Griffith (1861, p. 145) did not recognise it as a mine. The Memoir says that Griffith recorded bismuth here; but I cannot trace the reference.

Tynagh (Carhoon). 1″ 116. 6″ Galway 117 N.W. This was in Carhoon townland, and the lode is marked on the MS. 6″ Geological Survey map. Mem. 115 and 116, p. 39 (1865), notices remains of rather extensive works. Tynagh was unknown to Griffith. Kinahan (p. 12) speaks of Tynagh (Carhoon) as a very ancient silver mine. It is named in the Lists of Mines 1862-5.

Caherglassaun (*Caherglassen, Caherglassan, Cahenglassaur*). 1″ 115. 6″ Galway 122 N.E. Mem. 115 and 116, p. 38 (1865), describes this lode, with ore in bunches in calcite and no definite vein. Bournonite and traces of copper ores occur. It was worked about 1861, but was very wet, since the workings were below sea-level. Griffith (1861, p. 146) says it was mined by the Connemara Mining Company. The Lists of Mines 1862-5 give F. M. S. Taylor as proprietor.

The Lead Mines of Clare.

The Carboniferous Limestone that plays a large part in the structure of northern and central Clare is traversed by numerous veins bearing galena. The ore is usually argentiferous, and is associated in many cases with blende, and at Kilbreckan with antimony sulphide. Two areas stand out conspicuously, the Burren country of the north and the region east of Ennis and extending towards Tulla. Kinahan (pp. 13 and 14) calls the mines in the former district the Ballyvaughan Mines, and in the latter the Quin and Tulla Mines. The northern group is also known as the Burren Mines, and these were more generally worked in ancient times. The southern group was mainly developed after 1833, and we are indebted to Mr. E. R. Woakes, A.R.S.M., of Messrs. John Taylor and Sons, for access to some of the early Reports of the Clare Mines Adventurers who opened up the district. The names of the mines in this district are usually well indicated on the engraved Ordnance Survey 1″ sheet 133, which was used as the basis for the Geological Survey map in 1861.

The Burren Mines were reported on by Mr. A. E. I. M. Russell to the Department for the development of Mineral Resources in 1917.

Ailwee. 1″ 114. 6″ Clare 6 N.W. Little is known about this mine, marked as an old silver mine on the engraved 6″ Ordnance Survey map, and lying high up on the edge of the limestone plateau of the Burren. It is in Ailwee townland, and is recognised by Kinahan (p. 13). My attention has been called to it by Mr. M. Comyn, K.C., who has interested himself much in the mines of the northern district.

Mogouhy. 1″ 123. 6″ Clare 10 N.W. This was in Mogouhy townland half a mile N.W. of Castletown, and was being worked in 1863 (Mem. 114, 122, and 123, p. 27, 1863). An adit ran in from the face of the scarp; the MS. 6″ geological map suggests that this was from the north of Castletown townland; a shaft was sunk upon it. No output is recorded.

Castletown. 1″ 123. 6″ Clare 10 N.W. and S.W. This mine was somewhat S.W. of Mogouhy. Kinahan (p. 75) says that records of a "silver mine" here go back to the reign of James I. Castletown is named as producing both lead and zinc ores in Min. Stat. Lists of Mines 1862-5, when worked by Col. McNamara. There is, however, a Castletown mine in 1″ 133 (see later), which makes it difficult to assign the records definitely. The M.C.I. Report for 1845 says that the Company annulled their lease at Castletown, searches having proved unsatisfactory; they place this mine in Co. Limerick, and hence it may be the southern of the Castletown mines of Clare. See below.

Glendree. 1″⅜ 124. 6″ Clare 27 N.E. Mem. 124 and 125, pp. 24 and 48 (1863), mentions an adit in this valley, some 5 miles west of Kildavin. Hely Dutton (Stat. Surv. Clare, p. 13, 1808) says that he discovered very rich lead ore in Glendree; this may have been worked at a later date. Griffith's Map and the Geological Survey maps give no mineral indication here.

Ballyvergin. 1″ 133. 6″ Clare 26 S.E. This has been already mentioned under Copper. It lies in Ballyvergin townland, 4½ miles W.N.W. of Tulla. Mem. 133, p. 34 (1862), describes the lode. It has a brief official record of 55 tons of lead in 1859 and 64 tons in 1860. Mem. 133 says that it was abandoned in 1861, after working had gone down to 10 fathoms. It remains in the Lists of Mines until 1868 (D. T. Macdonald proprietor).

Carrahin (Carrahan, *Carahan*). 1" 133. 6" Clare 34 N.E. This mine is marked in Lassana townland on the engraved 6" Ordnance Survey map. The spelling Carrahan is probably correct, since a townland of that name lies to the N.E. Mem. 133, p. 135 (1862) regards it merely as a trial, and its name (placed in Co. Cork) appears in Min. Stat. 1863 as raising only 11 tons of lead. But in 1879 it yielded 111 tons, valued at £1,524, when the metal was at about £10 a ton, and 322 tons of ore (240½ tons of lead) in 1880. There is no further record. It was latterly worked by the Carrahan Silver Lead Company.

Crowhill (Crow Hill, Knockaphreagaun). 1" 133. 6" Clare 34 N.E. This is one mile E. of the Carrahin mine. Mem. 133, p. 35, says that the lode was worked to 22 fathoms, and was abandoned after the ore had been all removed. This was before 1862.

Milltown (*Miltown*). 1" 133. 6" Clare 35 N.W. This was an ancient mine, 1½ miles W. of Tulla, and is engraved as a silver mine on the 6" map. The ore is scattered through a great lode of calcite, which was worked open to the day. Stewart (p. 24) noted the "crystal spar" as 40 yards wide. Zinc was extracted from the blende associated with the argentiferous galena. The Royal Irish Mining Co. opened up the older workings about 1826 (P. M. Taylor, "A short account of lead mines in the Co. Clare," G.S.D., vol. i., p. 385, 1838), raised 11 tons of ore, and stopped. Mr. John Taylor of London reopened the mine for the Clare Mines Adventurers in 1837. Forty tons of ore yielded 75 per cent. of lead, with 37 oz. of silver to the ton. Weaver (1838, p. 66), from information given him by J. Taylor, quotes even 50 to 60 oz. to the ton. From the first, J. Taylor was not very hopeful, owing to the scattered nature of the ore, and the mine was abandoned in 1838. In 1853 an expert was called in to report to the Proprietors, and he drew attention to the prospects offered by the blende, below which further lead ore might be expected (see under Zinc). Min. Stat. 1862 show an output of 5 cwt. of lead, and nearly 3 tons in 1864; but this seems to have been only a trifling revival, following on some working of the blende (see under Zinc). Griffith (1861, p. 141) says that Milltown was worked by the Bullion Mining Company. The Lists of Mines give it as under the Clare Mining Company from 1860-8. The site is marked as a disused silver mine on the 1" Ordnance Survey sheet of 1907.

Ballyhickey (*Ballyhicky*). 1" 133. 6" Clare 34 S.E. This was by far the most important of the Clare mines. J. Taylor, in a note to P. M. Taylor's paper (*op. cit.*, p. 386), describes the galena as occurring in huge veins of calcite. The ore gave 77 per cent. of lead, but only 15 oz. of silver to the ton. The main vein or bunch was 16 to 20 feet wide and almost pure galena. Blende and copper pyrites were associated. Ballyhickey is called a lead and silver mine on the engraved 6" map of the Ordnance Survey.

Attention was first called to the "bunch" by the occupier, who noticed blocks of galena when making drains. J. Taylor took up work here for the Adventurers in 1836, and the early promise of the mine was remarkable for the district (Reports by J. Taylor for 1836, etc.) 1,300 tons of ore were raised by 1838. Weaver (1838, p. 66) notes that the deepest working in 1837 was 13 fathoms, and that a steam-engine was being furnished. 2,500 tons of ore were shipped to the Dee from the neighbouring port of Clare (Clarecastle), then newly constructed on the Fergus, in three years or so of working. Taylor's report for 1839, however, opens with the following words: "The rich discoveries of lead in the County of Clare have been of short duration; they were very unusual and remarkable in their character, and, unfortunately, their decline has been nearly as rapid as their rise." In 1840 the return was about 40 tons of ore per month; but the ore-ground was regarded as nearly worn out. Hunt (1848) gives an output of 119 tons of ore (only 32 tons of lead) for 1845, and 83 tons (22 tons of lead) in 1846. The ore had evidently fallen off in quality, and only the name of the mine occurs in the records for 1847-9 and again in 1853. It is revived in the Lists of Mines from 1860 to 1865.

Kilbreckan (Monanoe ; *Kilbricken*). 1" 133. 6" Clare 34 S.W. This mine was in Kilbreckan townland, between Ennis and Quin, and 1½ miles east of Kilbreckanbeg House. It is marked as "Silver Mine (Disused)" on the 1" Ordnance Survey map of 1907. The ore was discovered by some men in cutting a drain through a bog. J. Taylor (Report to the Clare Mines Adventurers, 1836) says that the ore-body was much like a pipe, unusually rich, but not extending more than 2 fathoms in length. This was the first mine opened up by Taylor in the district; Weaver (1838, p. 66) says that this was in September, 1834. P. M. Taylor (*op. cit.* under Milltown, p. 386) tells us that the first ore shipped assayed for lead 70 per cent. and for silver 120 oz. to the ton. Water was a trouble here from the first. A steam pumping-engine was erected in 1837, and the quantity of silver in the ore made the enterprise very attractive. The first bunch was worked out by 1838, and another was met with in a 50-fathom level from the bottom of the 10-fathom shaft. Prospects of fresh ore were, however, not favourable in 1840.

Hunt (1848) records 102 tons of ore (66 of lead) for 1845; 38 tons (25 of lead) for 1846; and 96 tons (67 of lead) for 1847. Returns then cease until 1853, and nothing occurs beyond 1855, though Holdsworth (" Geol. etc. of Ireland," p. 69) seems to have thought that the mine was still doing well in 1857. The name occurs in the Lists of Mines down to 1865.

A plan in the Home Office series (R 56) has a note on it stating that the mine was abandoned in 1856. Mem. 133, p. 35 (1862), gives the thickness of the main lode as 1 to 3 feet. The richest ore was got in a shaft that reached 30 fathoms.

The presence of a joint sulphide of lead and antimony at Kilbreckan has been already mentioned under Antimony.

Castletown. 1" 133. 6" Clare 34 S.W. A more northern mine of this name lies, as already noted, in 1" 123. Griffith's Map shows that the southern one was near Castletown Lough, 3 miles east of Ennis. Near Castletown House, which lies at some distance N.E. of the lough, there was formerly a village styled Castletown. **Moyriesk** mine, one mile to the east of Castletown mine, is associated with it by Griffith (1861, p. 141). Mem. 133, p. 36, regards this as a mere indication of lead ore. The date of working of these two mines is unknown. J. Taylor made trials at Castletown in 1838, finding no ore (Report to Clare Mines Adventurers, 4th July, 1838); and this mine may be the Castletown that was abandoned by the M.C.I. in 1845. See Castletown, 1" 123, above.

Kilmurry (possibly *Kilmorie*). 1" 133. 6" Clare 43 S.W. " Silver Mine " is engraved in Kilmurry on the 6" Ordnance Survey map, and Mem. 133, p. 36 (1862), records a vein. Min. Stat. Lists of Mines 1868-74 name " Kilmorie " as a silver lead mine, and 4 tons of lead are recorded as raised in 1868 from Kilmorie in Tipperary. This may possibly refer to Kilmurry in Clare, since, as we have seen, counties are not very accurately recorded in these publications; but it is also worth noting that the Silvermines mines of Tipperary occur in a parish of Kilmore.

Garrykennedy. 1" 125 (extreme south). 6" Tipperary 13 S.E. This mine is near the shore of Lough Derg, west of Youghal Bay, and about 1½ miles N. of Portroe. The ore is in joints of the Silurian grits and shales, and thickens out into " bunches " in the shales, contrary to the general rule of ore-deposition. The lode has arisen along the plane of a reversed fault. Ancient workings have been found. The maximum depth explored is 13 fathoms. The mine is described in Mem. 134 (not 125), p. 42 (1861), and the last trials are said to have been made about 1855.

Ballyhurly. 1″ 134. 6″ Clare 29 S.W. This mine was in Silurian strata on the south coast of Scarriff Bay, Lough Derg, 4 miles east of Tomgraney. It appears on Griffith's Map, on the authority of the author's MS. notes (1861, p. 141), and is mentioned as a trial in Mem. 134, p. 34 (1861).

Clasagh. I have not traced this mine, which Stewart (p. 24) says was in the "barony of Tullo" and "worked many years ago by a company" (say about 1760) for lead ore with silver and antimony.

Lead Mines of the Silvermines District.

As the name of Silvermines village implies, this district has long been known for its production of argentiferous galena. We have already noted that some copper ore has been raised, and the mines will again be referred to both under Sulphur and Zinc. A line of faulting, probably an Armorican feature, has lowered the Carboniferous strata on the north side of the Old Red Sandstone and Silurian mass of Silvermines Mountains, and the mines extend along it from Silvermines to Shallee Cross. Roads, as shown on the 1″ geological sheet 134.

Kinahan (p. 100) suggests that the district was worked by the ancient Irish. Gerard Boate, in "Ireland's Naturall History," p. 141 (1652), describes a mine in the County of Tipperary, parish of Kilmore, "not far from the castle of Downallie," as found out not above forty years before (say 1610, as he wrote in 1645), and worked for lead and silver. This seems to have been the only lead mine actively at work in Boate's time in Ireland, and only two others were known to him in the country (see under Coney Island). He records 3 lbs. (48 oz.) of silver to the ton, and "also some Quicksilver, but not any Alome, Vitriol, or Antimony, that I could hear of." Lead was at that time worth £11 a ton at the mine, and £12 in Limerick. The English and Dutch (possibly Saxon) miners had worked it, "because the Irish having no skill at all in any of these things, had never been employed in this Mine otherwise than to digg it, and to doe other labours." Boate, we must remember, wrote four years after the bitter events of 1641, when almost all these strangers were killed, the mine-works ruined, and mining enterprises stopped throughout the west and north. He points out, however, without comment, that the trouble at Silvermines was due to an endeavour by Hugh O'Kennedy to win back lands wrested from his brother by the Crown.

A. B. Wynne ("The Mining District of Silvermines," G.S.D., vol. 8, p. 245, 1860) says that the Dunalley family worked the mines at different times since 1720. J. Rutty ("Nat. Hist. of Dublin," vol. 2, p. 141, 1772) records 80 oz.

of silver to the ton of lead from Silvermines, which was the richest proportion known to him from Irish mines. Wynne (*op. cit.*, p. 249) confirmed this in 1860, no doubt on the basis of an analysis by Apjohn in G.S.D., vol. 8, p. 159. Weaver (1819, p. 242) says that working was resumed at Gorteenadiha in 1801; but things were not a success when he visited the mine in 1807. He gives a concise and clear account of the mode of occurrence of the ores. In his later paper (1838, p. 60), he mentions four vertical veins ranging N. and S. at Shallee, composed of barytes with fine-grained galena very sparingly scattered. Shallee had been worked by open cuts without much result. Weaver remarks on the occurrence as unusual in Ireland, being in Old Red Sandstone. Kane does not add to our knowledge of the district. Kinahan (p. 100) records that the General Mining Company for Ireland worked the mines down to about 1870 (see Shallee below). This Company must not be confused with the M.C.I., which entered on the "sulphur ramp" south of Silvermines from 1840 to 1843. A. B. Wynne gives a very useful map of the mining area in connexion with his paper above cited (G.S.D., vol. 8, 1860). In preparing Mem. 134 (1861), he devoted much attention to the mines. The names used in this Memoir should be compared with those adopted more wisely by Griffith (1861, pp. 151-2) from the Ordnance Survey maps. The Silvermines district was reported on by Mr. H. J. Daly to the Department for the development of Mineral Resources in 1917.

On Griffith's Map (1855) the sites of the mines are accurately placed, and in his list (1861) we find that lead ore was actually raised from all those mines along the Silvermines "channel." From west to east, we have **Shallee West** and **East, Gorteenadiha, Garryard West,** **Gortshaneroe** (also known as **Ballynoe**), **Knockanroe** (S.W. of Silvermines), **Ballygowan South** (or **Silvermines**), immediately south of the village, **Garryard East** at the village, **Cooleen** a little to the N.E. All these are in 6" Tipperary 26. Mem. 134, pp. 44-5, includes a table showing the characters of the lodes and the minerals at many of the mines.

Lead-mining records exist in regard to the following:—

Shallee West and **East.** 1" 134. 6" Tipperary 26 S.W. The first record of output is in Hunt (1848), who gives 209 tons of ore (yielding 125 tons of lead) for 1847. The two Shallees gave 467 tons of lead and 12,000 oz. of silver in 1852. In later years the output became much smaller, ending with 2 tons 12 cwt. in 1874, when lead was at £13 to £19 per ton. From 1870 on, the mines were taken over from the General Mining Co. for Ireland by the Shallee Silver Lead Mining Co., which was in liquidation from 1874-7, and was succeeded by Mr. Charles Cummins.

The Shallees are placed in the county of Waterford in Min. Stat. for 1856 and many subsequent years. A MS.

report on mines and mining at Shallee was made by Mr. T. Hallissy for the Geological Survey in 1917.

Gorteenadiha (*Gurteenadyha, Gurtnadyne, Gortnadine*). 1" 134. 6" Tipperary 26 S.W. This mine is in the L. Carboniferous Shale on the north side of the great fault, and just east of the stream in Gowlaun Glen. It is an old copper-producing locality, and first appears in Min. Stat. as a lead mine in 1852. Fifty tons of lead were obtained in this year; 154 tons, with 3,280 oz. of silver, in 1853; and then greatly diminished quantities down to $14\frac{1}{2}$ tons of ore in 1860, when the record of output ceases. After appearing (under copper) in Co. Wicklow in Min. Stat. 1854 and other years, the mine is said to be in Waterford in 1856 and in the Appendix for 1861. It was worked for lead by the General Mining Co. for Ireland down to 1869, when it passed to the Shallee Silver Lead Mining Co. It is stated to have "stopped" in 1870, but the name appears in the Min. Stat. List of Mines down to 1874. See also under Copper.

Silvermines. 1" 134. 6" Tipperary 26 S.E. Min. Stat. 1865, record under this name an output of $8\frac{1}{4}$ tons of lead in 1865; 1 ton 2 cwt. in 1866; and 27 tons with 1,111 oz. of silver in 1868. The name appears in 1870 and later years, and may represent Griffith's Ballygowan South (*Ballygown*) (see above).

NOTE.—The orpiment in barytes described by J. Apjohn (G.S.D., vol. 8, p. 243, 1860) came from Ballynoe (see Wynne, *ibid.*, p. 246).

Ballysteen (*Ballystein*). 1" 143. 6" Limerick 11 N.W. Ballysteen Mine was in Ballycanauna townland, near Ballysteen House and the Shannon shore north of Askeaton. Mem. 143, p. 34 (1860), says that it was on a rich pocket of argentiferous galena in a lode that ran E. and W., but was worked out and closed before 1860. As a matter of fact, it was abandoned when Weaver wrote in 1836 (1838, p. 65); he says that one shaft reached 17 fathoms, and there were shallow works 80 fathoms long. The site is marked as a silver mine on the engraved 6" Ordnance Survey map.

Ballinvirick (**Ballintredida**, *Ballintreddedy*). 1" 143. 6" Limerick 20 N.E. Ballinvirick is in Ballintredida townland, and often bore its name. It lay between Askeaton and Adare. Weaver (1838, p. 65) records a calcite vein 3 or 4 or perhaps more feet wide as then recently discovered. It had "only been touched on, the proprietor delaying to make adequate researches." The minerals were an interesting group, "antimonial galena, earthy carbonate of lead and oxide of antimony, grey copper ore, antimonial copper ore, blue and green carbonates of copper, and blende." Neither the lode nor the mine were traced with certainty by the Geological Survey, and the lode is not marked on the 1″ map. See Mem. 143, p. 35 (1860). On the MS. 6″ map what "seems to be the shaft of an old mine" is noted.

Ballinglanna. 1″ 151. 6″ Kerry 9 S.W. A mine was formerly worked here north of the village of Causeway, inland from the Old Red Sandstone mass of Kerry Head. It was known to Griffith (Map and 1861, p. 147); but the Geological Survey (Mem. 150 and 151, p. 16, 1859) found only lumps of galena in a stream flowing through the townland.

Mahoonagh. 1″ 152. 6″ Limerick 36 S.E. Stewart (p. 100) speaks of "former great works" at the bottom of a 14 feet shaft at "Mahonagh," which is S.E. of Newcastle. There was a small vein of rich lead ore; date of working unknown. Griffith (1861, p. 148) knew of this old mine. The 6″ Geological Survey map gives no indication.

Cloghatrida. 1″ 153. 6″ Limerick 20 S.W. "Lead Mines" appear here on the engraved 6″ Ordnance Survey map. The lode forms part of the group that gave rise to the Ballinvirick and Ballysteen mines further north; it runs E. and W., and was also worked in **Ballingarane.** Mem. 153, p. 29 (1861), says that all the ore was taken out between the two places. The gangue was calcite and ferruginous dolomite. The zinc symbol has been inserted at Cloghatrida on the Mineral Map, owing to the occurrence of blende; but it might equally have been inserted in other places where unutilised zinc ore occurs. Cloghatrida was abandoned, apparently through water troubles, before 1857.

Mayne (*Main*). 1″ 153. 6″ Limerick 45 N.W. Stewart (p. 99) records several small pits in this townland on lead veins. The MS. 6″ Geological Survey map gives no indication.

Oola Mines (Oolahills). 1″ 154. 6″ Limerick 24 S.E. and 25 S.W. Mem. 154, p. 28 (1861), fully describes the lode that runs here E. and W., about ½ mile N.E. of Oola railway station, and is traceable eastward for ¾ mile. Several shafts were opened on it; but in 1859 working was confined to the western end, and the mine was closed, for a time at any rate, before 1861. Griffith (1861, p. 148) says that the proprietors were the Oola Silver, Lead and Copper Mining Company. Min. Stat. Lists of Mines give H. Wadge as proprietor in 1865, and W. Sunderland and Co. from 1876-9. Copper pyrites was raised as well as galena (see under Copper); the gangue was barytes, which is here plentiful in vertical veins. The galena extracted gave 29 oz. of sliver to the ton.

Knockadrina (Flood Hall). 1″ 157. 6″ Kilkenny 27 S.E. According to Kinahan (pp. 20 and 89), this mine was very ancient and known for its silver. Tighe (Stat. Surv. Kilkenny, p. 88, 1802) gives it as working when he wrote, under Mr.

Wyse of Waterford. Weaver (1819, p. 282) and Griffith ("Mines of Leinster," p. 29, 1828) knew it by reputation. Kane (1845, p. 210) says that in respect of silver the ore was very rich, but working had ceased by his time. There being no record of output, Kinahan's remark (p. 89) that successful mining of lead and silver took place here " recently "—that is, near 1887—seems doubtful.

Caim and Ballyhighland (*Caime; Ballyhiland.*) 1" 158. 6" Wexford 19 S.W. The mine is in Ordovician strata in Ballyhighland townland, 5 miles west of Enniscorthy. Caim townland is adjacent on the east. Weaver (1819, p. 219) mentions the discovery of ore here and the opening of the mine as occurring a few years before 1818. The gangue was quartz, and iron and copper pyrites occurred with galena, cerussite, and blende. Twenty-four tons of copper pyrites and "a few hundred" of galena had been shipped. The shaft then reached 24 fathoms, but the mine was abandoned for lack of machinery. The M.C.I. noted the old shafts, and took it up about 1836 (Rep. 1837 i), working it at a small loss and in a very small way. It was doing well by the end of 1838, and in both 1840 and 1843 some 270 tons of lead ore were raised in six months. Kane (1844, p. 198) gives 505 tons of ore for 1842. He found 130 persons employed here by the M.C.I., but evidently regarded the fortune of the mine as speculative. The galena gave 75 per cent. of lead, and was smelted at Ballycorus. The M.C.I. (Rep. 1844 i) reported that a newly found lode proved to be "again cut off," and work was in consequence suspended. Litigation occurred as to further extensions; Smyth ("Mines of Wicklow and Wexford," p. 398, 1853) regarded Caim as practically abandoned before 1846. The levels below the opencast were flooded at that date. Some work (M.C.I. Reports) was still, however, carried on at a small loss down to 1855, and the mine is named in the Min. Stat. Lists from 1860-5.

W. W. Smyth placed in the Mining Record Office (Home Office series, R 301; the date of deposit, 1841, should, perhaps read 1847) a tracing after a map by Brockenshar made "previous to being knocked in 1845." This copy is dated June, 1846; but, as we have seen, some working went on to a later date. The drawing shows a large opencast to the 13 fathom level, and the deep level at 57 fathoms. Smyth ("Mines of Wicklow and Wexford," p. 397, 1853) describes the lode, and speaks of a 67 fathom level, where the mass of galena, which was 12 feet wide above the 47 fathom level, had narrowed down to 5 feet. He mentions (p. 398) 15 oz. of silver to the ton.

Barrystown (*Barristown, Barrastown, Banistown, Barretstown*). 1" 169. 6" Wexford 45 N.E. The name appears in many forms; "Barretstown" is apparently peculiar to

Kinahan. The mine is in Barrystown townland, on the east side of the narrow head of the Bannow estuary. R. Fraser (Stat. Surv. Wexford, p. 13, 1807) says that it was worked thirty or forty years before 1807 by the owner of the property, Mr. Ogle, but with no profit. On p. 16 Fraser mentions, in a very unsatisfying way, a manuscript that he had seen in the Archbishop's library in Lambeth, where a silver mint in the County of Wexford is attributed to the Danes. This may be the basis of other statements connecting a mint with Barrystown. Griffith ("Mines of Leinster," p. 22, 1828) knew the workings as ancient, and that they were attributed to the Danes. Kinahan (p. 479) gives an interesting history of mining in this area in Tudor times, including the adjacent **Clonmines** mine, the site of which is now lost.

W. W. Smyth ("Mines of Wicklow and Wexford," p. 398, 1853) says that the richness of the ore in silver, 60 to 70 oz. of silver to the ton, attracted attention to the mine in 1846. Hunt (1848) has a record of 22 tons of ore, yielding 14 tons of lead, in 1845; 250 tons of ore were raised in 1846, and 301 in 1847. The mine is erroneously placed in Co. Waterford. Returns are given in Min. Stat. for 1848 and 1850, but not later; when Smyth wrote (1853) working had ceased. The M.C.I. (Report for 1857) made searches here, but found next year that they afforded "no reasonable prospect of success."

Smyth states that the lode is very variable, but about 3 feet wide, the galena and blende being in a gangue of quartz and siderite. The lode was, as far as he could see, broken by a series of reversed faults. The shaft reached 18 fathoms, and a pumping-engine had been put up in 1846 for more extended operations. Prospecting has recently been resumed here through the energy of Col. E. A. Johnson (His Excellency Johnson Pasha).

Annagh. 1" 173. 6" Kerry 47 N.W. The mine was in Lower Carboniferous Limestone, close to the road leaving Castlemaine on the north side for Tralee. The best account of it is in Weaver (1838, p. 64), who gives a plan and section. He says that it was discovered in 1788, worked for three years and then abandoned. R. E. Raspe, the German expert, (see Muckross), reported on it in October, 1793; he says that it had yielded 400 tons of argentiferous galena, from workings down to 12 fathoms. In 1825, the mine was again taken up; but it was found that the great mass of calcite and quartz in which the ore had occurred thinned out below, so that only 9 tons or so of lead ore and a few tons of blende were raised by these resumed operations. The lead yielded 43 oz. 11 dwts. of silver to the ton. Plans of Annagh, dated 1826 and 1828, are in the Home Office series (R 286).

Meanus. 1" 173. 6" Kerry 47 N.W. Nothing seems definitely known about this mine, of which Griffith (1861

p. 147) names the Resident Director. Griffith's Map shows it correctly, near Annagh, and on the opposite side of the highroad to Tralee. Mem. 173, p. 25 (1861), doubts Griffith, and points out that a Meanus townland is in 6" 57; but the name is also that of a townland immediately north of Castlemaine. The MS. 6" Geological Survey map shows a shaft here in Meanus, on the easterly continuation of the Annagh lode. Du Noyer, who wrote the Memoir, was evidently not aware of this.

Annestown. 1" 178. 6" Waterford 25 S.E. At this spot, east of the famous Knockmahon copper-mining district, in Ordovician strata on Dunahrattan Bay, the M.C.I. prospected for lead ore in 1837 (Report 1837 ii), but concluded their trials without satisfaction in 1838. Searches for copper ore were made in 1869.

Stewart (p. 134) in 1800 mentions "in a high cliff on the shore," 2 miles S. from the Bunmahon copper mines, a lead mine "very rich in silver, but poorly worked, by Mr. Wyse, of Waterford." Two miles south of Bunmahon of course bring one well out at sea; but it is possible that the reference is to something near Annestown.

Cloontoo (*Clontua*). 1" 184. 6" Kerry 93 N.E. Griffith 1861, p. 147, associates Ardtully and Cloontoo as alternative names for the Ardtully copper mine (see under Copper) east of Kenmare. In his Map he indicated copper only at Ardtully, omitting the name Cloontoo. As W. B. Wright shows in his table of the lodes, which we have given under Copper, Cloontoo, lying close to the west of Ardtully, produced both copper pyrites and galena, and Weaver (1838, p. 28) clearly regards it as a lead mine. Haughton (G.S.D., vol. 6, p. 213, 1855) does not mention Cloontoo separately.

Shanagarry (Caher West, *West Cahir*). 1" 184. 6" 93 N.E. This lode is lower down the Roughty valley than Cloontoo, and near the village of Cleady. Trials had been made on it in Weaver's time (1838, p. 28) to a depth of 46 feet. This was probably the mine that Kane (1845, p. 209) mentions as working at Kenmare in 1844. It is named as a silver-lead mine in the Lists of Mines for 1879. Haughton (G.S.D., vol. 6, p. 213, 1855) describes the lode as containing a pure argentiferous galena in the lower levels. Mem. 184, p. 37 (1859), gives the depth of working as 40 fathoms. The lode is in the limestone of the Kenmare syncline. Still further west, there was another series of pits for lead ore in **Killowen,** only a mile east of Kenmare. Haughton (*op. cit.*) regarded the Killowen lode as important.

Ardmore (Dysert). 1" 188. 6" Waterford 40 N.E. Stewart (p. 135) mentions a mine, with a shaft at work in 1800, as near Ardmore church "opposite to Youghal." Youghal is 5½ miles to the west. There was a small vein of galena.

Griffith marks a lead mine here as "Dysert" on his Map of 1855.

Ringabella. 1" 195. 6" Cork 99 S.E. A lead mine was worked in the early part of the nineteenth century in Carboniferous Slate just south of the village at the head of Ringabella Bay. Weaver (1838, p. 25) says that it was a "bed" a few inches up to 1½ feet wide. In his time the trial had reached 20 fathoms. Iron and copper pyrites occur, the former being more abundant. Kane (1845, p. 209) hints that the various workings and frequent abandonment of this mine do not speak much in its favour. Griffith (1861, p. 143) says that the ore was argentiferous. It seems to have escaped mention in Mem. 187, 195 and 196.

Cahermore (Kilkinnikin West). 1" 198. 6" Cork 127 S.W. The mine was just south of Cahermore village, and 3 miles south of the Allihies copper mines. Mem. 197, 198, p. 29, says that the shaft was still open about 1860.

Killoveenoge (Bantry Lead Mine; *Killovenogue*; *Killevenoque*) and **Rooska East.** 1" 199. The name Rooska on the Mineral Map indicates a lode running E. and W. in slates from the townland of Killoveenoge through Rooska West, the next townland on the east, and into Rooska East. Griffith on his Map and in 1861 p. 142 recognised two mines here, Killoveenoge and Rooska East, about one mile apart. The ore was argentiferous galena. Killoveenoge was at work about 1863 (Mem. 192 and 199, p. 47, 1864), with two shafts. The lode was 3¼ feet wide. Killoveenoge is marked "Silver Mine" on the engraved 6" Ordnance Survey map; the sign for lead on Sheet 199 of the Geological Survey lies in Rooska West. From 1875-9 it seems to have raised barytes (Lists of Mines).

The outputs quoted in Min. Stat. from **Bantry** are, no doubt, from the "Bantry Lead Mine, Killevenoque," of the Lists of Mines 1863-78. They are :—1849, 14 tons of lead ; 1850, 13 tons 9 cwt. ; 1852, 10 tons.

Gortacloona. 1" 199. 6" Cork 118 S.W. This lode is crossed by the high road from Bantry to Ballydehob, 1¾ miles due S. of Bantry. The Hollyhill copper lode lies a little north of it. The lode marked on the 1" Geological Survey sheet is Gortacloona. It seems to have escaped mention in both of the two Memoirs dealing with Sheet 199. It appears as one of the two Hollyhill Mines in Griffith 1861, p. 142. *Gartydona* appears as a Cork lead mine (Zohrab Holmes and Co.) in Min. Stat. Lists of Mines 1862-5, and may perhaps be Gortacloona.

Ballydehob. 1" 199. 6" Cork 140 N.W. See under Copper, for which this mine was mainly worked. Min. Stat., however, record 32 tons of lead as extracted from 50 tons of lead ore in 1856.

Lisduff. 1" 201. 6" Cork 144 N.W. The MS. 6" map of the Geological Survey records several tons of lead as raised from veins in this townland west of Clonakilty Bay. No reference to a mine.

Duneen Bay (Duneen; *Doneen* ; **Muckruss Head).** 1" 201. 6" Cork 144 N.E. This mine has been dealt with under Barytes, which is its principal product, and also under Copper. Weaver (1838, p. 25) noted lead ore in the old workings, which were said to have been carried to 30 or 40 fathoms, and it was certainly regarded as a mine both of lead and copper (Griffith, Map and 1861, p. 142). Mem. 194, 201, 202, p. 25 (1862), has only a casual mention of the workings in the "glossy bluish-gray slates" of the Lower Carboniferous series.

Boulysallagh. 1" 204. 6" Cork 147 S.E. This mine, north of Crookhaven, has been mentioned under Copper; but it also raised some argentiferous galena (Griffith, Map and 1861, p. 142).

From 1860-5, a **Hibernian** lead mine is cited in Min. Stat., but without a county reference. An untraced *Lansdown* mine in Co. Kerry is named in the same years.

CHAPTER XI.

MANGANESE.

Calliagh. 1" 58. 6" Monaghan 13 S.W. References to the literature on the manganiferous iron ore in this townland will be found under Iron. Adeney's analysis, quoted in Mem. 58, ed. of 1914, gives 6·24 per cent. of manganese peroxide, and samples examined by him showed a range from 5·9 to 7·55 per cent. Kinahan (p. 480) erroneously quotes these figures as "manganese." His townland of "Tattin Heive" should be Tattintlieve.

Westport. 1" 74 and 84. Bog-manganese ore occurs in this district, and probably accounts for the reference in J. McParlan, Stat. Surv. Mayo, p. 20 (1802), to manganese from the mountains of "Glanmore," 4 miles S. of Westport, which had been in his time shipped to England. "Glanmore" seems to be Lanmore townland, in 6" Mayo 98 N.W., where slate has been quarried. A large bog stretches to the south.

Sutton. 1" 112. 6" Dublin 15 S.E. Griffith (1861, p. 145) gives manganese ore as raised at Sutton, on the south side of the Howth peninsula. Du Noyer, in Mem. 102 and 112, p. 70 (1875), says that manganese was at one time found abundantly here in the Carboniferous dolomite, "but the supply appears to have been worked out." The brown-black colour of one or other of the oxides stains the walls of the cavities in the rock at present exposed upon the shore. Prof. H. J. Seymour (Mem. 112, p. 130, 1903) says that the ore was manganite, and probably occurred in a crush-breccia.

Kilbride (Cloghleagh and **Knockatillane).** 1" 120. 6" Wicklow 5 N.E. and 6 N.W. See under Iron.

Feakle. 1" 124. The bog-manganese ore of this district in the north-east of Co. Clare is referred to in Mem. 124 and 125, p. 48 (1863). Kinahan here states that earthy dialogite ($MnCO_3$) is common in the drift of the district. Like the deposits of some other localities, it may be worthy of further consideration. The name is on the Mineral Map just below that of the Kildavin mine.

Ballard. 1" 130. 6" Wicklow 30 S.E. and 31 S.W. This lode has been described under Iron. The upper part of the magnetite lode becomes limonitic and rich in manganese oxide. C. R. C. Tichborne's analyses (R.G.S.I., vol. 4, p. 269) were of the magnetite, which shows only a trace of manganese oxide, and of a more impure variety with 22·03 per cent. of haematite and 2·05 of manganese oxide. The analysis of the manganiferous portion of the lode, furnished by Mr. W. H. Pearson (see under Iron), gives manganese peroxide 18·08 and manganese protoxide 6·18 per cent.

Glandore Mines. 1" 200. 6" Cork 142 N.E. and 143 N.W. These old mines, **Leap** and others, on the lode in Aghatubridbeg and its continuation, have been mentioned under Iron. They afford the only adequately recorded case of manganese mining in Ireland. Kane (1844, p. 211) mentions "considerable quantities of the ore" of manganese as having been raised at Glandore before 1844. Min. Stat. first mention the lode as an iron mine in 1860, and give 60 tons 13 cwts. of brown haematite, value £35, as raised at Glandore, in that year. Griffith (1861) gives Aghatubrid, Glandore, as having been worked for manganese and copper; but Kinahan, in Mem. 200, etc., p. 25 (1861), says that manganese working was discontinued when he visited the mine (probably before 1860). He remarks that iron ore replaces the manganese ore in depth, and in his "Economic Geology," p. 80, suggests that copper ore may exist below. Wyley (Mem. 200, etc., p. 25) described the vein in a note in 1854 as "very solid and massive beds of binoxide of manganese, seldom less than 6 feet in thickness, occuring solid and in stalactitic and botryodal masses like haematite. Associated with it are heavy spar, haematite, and quartz, but no copper ore."

Kinahan ("Valleys and their relation to fissures, fractures, etc.," p. 27) says that the fault-rock in the fissure at Glandore was formed before the ores were deposited, since the haematite lode crosses and sends veins into it, while veins of manganese ores cut them both.

Glandore appears in Min. Stat. Lists of Mines 1862 to 1870, as a manganese mine worked by Tonkin & Co., but no output is given. All manganese mines suffered in 1863 from Spanish competition. While the mines in Devon and Cornwall revived, no further mining seems to have been done at Glandore till 1876, when Messrs. Foster and Willis worked the lode.

YEAR	OUTPUT: MANGANESE ORE	VALUE
1876	90 tons	£360
1877	50 ,,	150
1878	50 ,,	100
1879	40 ,,	80
1880	100 ,,	175
1881	250 ,,	435
1882 Worked by Messrs. Gordon & Co., 3 Westminster Chambers, London	No output recorded	
1883–1907	No record	
1908 Leap, Glandore, was worked by Liverpool Manganese Co.	156 tons	£128
1909	70 ,,	35

The Glandore Mines were reported on by Mr. H. J. Daly to the Department for the development of Mineral Resources in 1917.

CHAPTER XII.

MOLYBDENUM.

Molybdenite, the mineral form of molybdenum sulphide, MoS_2, may occur in the granitoid rocks of western Ireland in various places; S. Haughton, for instance, mentions an instance at Garvary Wood, Castle Caldwell (1″ 32) (Quart. Journ. Geol. Soc. London, vol. 18, p. 417, 1862). The only localities where molybdenite is yet known to be conspicuous are those marked Mo on the Mineral Map, and at none of these has commercial working been undertaken.

Inishdooey. 1″ 3. 6″ Donegal 14 N.E. An islet between Tory Island and the mainland. Mem. 3, 4, etc., p. 116, 1891. The molybdenite is said to be in a vein of dark quartz that traverses the limestone and quartzite of the pre-Cambrian series.

Lough Anure. 1″ 15 (extreme north). 6″ Donegal 41 S.E. The locality is a little east of the high road from Dunglow to Gweedore, and 3½ miles north-east of Dunglow, in the townland of Lough Anure. In granite.

Lough Lara. 1″ 15 (extreme south). 6″ Donegal 74 N.E. The spot is a little north-east of the road from Glenties to Maas, and 3 miles from Glenties, in the townland of Letterilly. In granite.

Murvey. 1″ 103. 6″ Galway 62 N.E. The locality is on the south side of Lough Namanawaun, south of the road from Roundstone to Clifden, in the townland of Murvey, in granite. The occurrence was reported on in 1917 by Mr. H. J. Daly to the Department for the development of Mineral Resources.

CHAPTER XIII.

NICKEL.

Though no important occurrences of nickel ore are known in Ireland, the frequency of pyrrhotine, associated with copper pyrites and iron pyrites, in the county of Galway west of Oughterard renders prospecting still desirable. See Mem. 93 and 94, pp. 163 and 164 (1878), and Kin., p. 37. The so-called **Maumwee** lode (1″ 94. 6″ Galway 38 N.E.) is described in the above-mentioned Memoir, p. 163, and was worked by H. Hodgson[*] in **Teernakill South** in 1860. A shaft 8 fatnoms deep was sunk on it. S. Haughton ("On the occurrence of nickeliferous Magnetic Pyrites from Tiernakill near Maum," G.S.D., vol. 9, p. 1, 1861) analysed a sample, which showed 60·41 of iron, 39·06 of sulphur, and 0·07 of nickel. A. Gages (Mem. 93 and 94, p. 163) verified the occurrence of pyrrhotine. See also under Sulphur.

---

[*] Formerly of Ballyraine, Co. Wicklow. See Smyth, "Mines of Wicklow," p. 376 (1853).

CHAPTER XIV.

ROCK-SALT.

Rock-salt in very considerable quantity occurs in the Triassic strata of north-eastern Ireland, as the product of desiccating lakes. Its discovery was due to borings being made in the hope of reaching coal. The main beds of salt are struck at about 600 feet below the surface. The total output from Co. Antrim has been maintained for many years at about 35,000 tons, reaching 50,000 tons, with unrecorded brine, in 1912.

The following notes have been largely aided by a Report made by Mr. T. Hallissy for the Geological Survey in 1919.

Magheramorne. 1" 21. 6" Antrim 41 S.W. Mr. John Vint of Belfast, a Director of the Salt and Alkali Co., Ltd., has kindly supplied the Geological Survey with records of five bores made in Triassic beds at Magheramorne in the townland of Ballylig (see below) between 1891 and 1914. Frequent alternations of rock-salt, marl and gypsum were proved, the most promising bed of salt apparently being that encountered at 980 feet in the diamond bore of 1914. Brine has been pumped for use in the Salt and Alkali Works from a depth of 318 feet from a shaft sunk 600 feet W. of the site of this diamond-bore.

Ballylig. 1" 21. 6" Antrim 41 S.W. Mem. 21 etc., p. 11 (1876), says that "Mr. Irving when sinking (or boring) for coal at Ballylig (printed *Ballybig*) came upon rock-salt, but did not persevere in his investigations." This appears to have been the first observation of salt-beds in the Irish area, but the date is not quoted. The recent borings at Magheramorne, Ballylig, have not traversed any continuous mass of this kind.

Ballyedward. 1" 21. 6" Antrim 47 N.W. J. B. Doyle (G.S.D., vol. 5, p. 254, 1853) says that P. McGarel, in boring for coal in 1839 in "Ballyedmond," reached rock-salt at 150 feet, the main bed being 24 feet thick. The bore was stopped at 174 feet. On the MS. 6" Geological Survey sheet a "Salt Hole" is marked about 100 yards S. of the boundary of Ballyedward townland, in Aldfreck, with a note by Du Noyer, "Salt said to have been discovered in the bottom of this small hollow."

About one mile to the S.E. of this Salt Hole, and N.E. of Ballycarry railway station, is **Redhall,** where Doyle (*op. cit.*, p. 235) says that "a new salt mine" was discovered about 1853. Doyle remarks, however, "but coal is the great object to be obtained." It may be noted that a "mine" in early usage often means merely a mass of mineral.

Duncrue. 1″ 29. 6″ Antrim 52 N.E. In Middle Divis townland, 1¾ mile N.W. of Carrickfergus, J. B. Doyle ("Notes on the salt mine of Duncrue," G.S.D., vol. 5, p. 232, 1853) describes the finding here of thick beds of salt in the course of searches for coal made by the Marquess of Downshire in 1853. A. Miscampbell ("The Salt Industry of Carrickfergus," Trans. Fed. Institution of Mining Engineers, vol. 7, p. 546, 1894) says that the operations began as far back as 1845, and that "a circular shaft was sunk, 9 feet in diameter, lined with brick to a depth of 750 feet, and from that point a boring was made a further 500 feet, but without finding coal." The rock-salt was encountered at 550 feet, as a seam "about 120 feet in thickness, with a couple of thin seams of mixed salt and marl interlaid." From Doyle's more detailed account, it seems that the salt was worked in 1853, soon after its discovery, which throws doubt on Miscampbell's early date, The Belfast Mining Co. took up the works (Miscampbell, p. 546). and made a shaft 1½ mile nearer to Carrickfergus harbour, and another one mile to the east of the original sinking, but without success. Prof. E. Hull, in the discussion on Miscampbell's paper, pointed out that these trials were on strata below those containing the salt zones, the dip being inward from the shore.

Two shafts were then sunk by the Company about 360 feet S.W. from the original sinking and 36 feet apart. The section was:—

	FEET
Boulder clay	50
Marl with gypsum	416
Rock-salt	11
Marl and rock-salt	5
Rock-salt	74
Marl and rock-salt	5
Rock-salt	37
	598

These figures are very close to those given by Doyle for the original sinking, and in a combined form in Mem. 21, etc., p. 10 (1876). After 50 feet of boulder-clay and 500 feet of red marls with thin bands of gypsum, a bed of rock-salt 15 feet thick was reached. The second bed is 88 feet and the third 39 feet thick. Kinahan (p. 56) makes a distinction here between beds of "pure salt" and "rock-salt," which is obscure, since he uses the term "pure salt" in two senses.

The Duncrue deposits may form a dome rather than continuous beds in the general dip, since they die out on two sides of the mine. A new shaft was put down here by the Belfast Salt Mining Co. in 1872, to a depth of 485 feet 6 ins., proving:—

ROCK-SALT

	FT.	INS.
Boulder-clay	60	0
Marls	330	0
Rock-salt and marl . .	20	0
Rock-salt	67	6
Brine	8	0

The Duncrue rock-salt yields some 96 per cent. of sodium chloride.

The Duncrue mine was developed rapidly after the discovery of the salt. Min. Stat. record 20,000 tons as raised in 1855, and some 27,000 in 1857. In consequence of the large excavations becoming unsafe, mines were opened up by the Belfast Salt Mining Company in 1870 in the neighbourhood, and the records are often combined; when Duncrue alone is made responsible for 16,845 tons in 1881, it seems probable that French Park is the actual source. A Home Office plan (No. 1933) gives the mine as abandoned in 1886; but there is a record of 7,144 tons, perhaps representing material already raised, as sold from Duncrue in 1887.

Though Duncrue became practically closed in 1870, brine was pumped for a time from its workings into the adjacent French Park mine. J. Rigby ("Outburst from Duncrue old Rock-salt mine after being tapped for boring," Trans. Manchester Geol. and Mining Soc., vol. 28, p. 565, 1905) gives an interesting account of an explosive outburst that occurred at the surface in 1899, through the air-compression caused by subsidence. A similar case is recorded from Marston, Northwich, in July, 1920.

French Park. 1″ 29. 6″ Antrim 52 N.E. This shaft was put down by the Belfast Salt Mining Co. in 1870, 500 feet N. of the old Duncrue pit, when excavation had rendered the latter mine unsafe. Its independent output is given in Min. Stat. as 17,430 tons in 1879. It was working under the Salt Union in 1918.

Maiden Mount. 1″ 29. 6″ Antrim 52 N.E. In Middle Divis, and ⅓ mile N. of Duncrue. The separate output is given as 14,100 tons in 1877; 17,394 in 1878; and later as about 13,000 tons per annum. The mine was worked, with French Park, by the Salt Union. The beds are:—

	FT.
Boulder-clay	86
Marl	686
Rock-salt and marl (roof of mine)	45
Rock-salt (worked) . . .	47
Rock-salt and marl . . .	19
Rock-salt (worked) . . .	30

913

A sump 10 feet deeper was in rock-salt and marl.

Burleigh Hill. 6″ Antrim 52 N.E. A shaft was put down about 1880 in the demesne of Burleigh Hill, south of the house, and ¼ mile E. of Maiden Mount. The beds are :—

	FT.
Boulder-clay	112
Marl	717
Rock-salt (worked)	24
	853

The mine was worked by the Salt Union, and was closed about 1900. Miscampbell (*op. cit.*, p. 550) says that the seam of rock-salt is very thin and of small extent. It is "cut off on two sides of the mine, and altogether appears to be a mere pot."

Eden. 6″ Antrim 53 N.W. A salt spring, marked on the engraved 6″ sheet and known to Doyle (*op. cit.*, p. 235), occurs about 700 yards N.W. of Eden village, which is served by Kilroot railway station. Miscampbell (*op. cit.*, p. 550) regards this spring as the centre of a basin of rock-salt of very small area. The Eden shaft was put down about 1890, between the spring and the village A bed of rock-salt 73 feet thick was reached beneath 227 feet of marls; 25 feet are allowed to remain as a roof, and 48 feet are mined, the bottom being reached in only one part of the mine. The phenomena of Duncrue seem to be repeated here, since a bore about 50 feet east, 451 feet 5 ins. deep, traversed numerous beds of good salt up to 20 feet in thickness, but not the main bed in which the great chamber of the mine is excavated at Eden. Miscampbell's opinion thus seems justified. Eden was one of the Salt Union mines, and it appears that it is now worked by the Chemical Salt Co., of Glasgow, under the name of the **Tennant** mine.

New Mine. 6″ Antrim 53 N.W. This is a mine of the International Salt Co., of Carrickfergus (formerly of the Salt Mines Syndicate), a little N.N.W. of the Eden shaft. Two beds of rock-salt occur, 36 feet and 80 feet thick respectively. The Manager informs us that the depth of the shafts is 550 feet. Working in 1918.

Downshire. 6″ Antrim 53 N.W. This is the mine of the Carrickfergus Salt Works Co., close to and N.W. of the Eden salt-spring. Working in 1918. It is said to reach 550 feet.

CHAPTER XV.

STEATITE.

Kinahan, p. 55, rightly remarks that various massive minerals, including pyrophyllite, have passed under the common name of steatite, which should be reserved for the massive form of the hydrous magnesium silicate, talc. It must not be concluded, however, that Kinahan's distinctions, made in his list of numerous localities, rest upon chemical observations.

Short of complete analysis, colourless specimens of pyrophyllite and steatite may be distinguished by moistening with cobalt nitrate solution and heating strongly in the oxidising flame of the blowpipe. Pyrophyllite gives the blue alumina reaction, and steatite the pale pink due to magnesia; but lack of opportunity seems to have prevented Kinahan from utilising a test of which he was aware ("Notes on Mining in Ireland," Trans. Inst. Mining Engin., Newcastle-on-Tyne, 1904, reply to discussion).

Carrowtrasna (Gartan). 1″ 10. 6″ Donegal 44 S.W. Steatite occurs here in workable quantities, in the general line of strike of the Dalradian rocks, where they are invaded by the Glendowan granite west of Lough Akibbon. The locality is ¼ mile N. of the well known Abbey Church associated with St. Columba, and 3 miles N.W. of Church Hill railway-station. Mem. 3, 4, etc., p. 57 (1891), mentions the "camstone" here, and on p. 117 it is said that works were carried on by Mr. Dudeworth from 1860-70. The steatite bed is 5 feet thick; it is, by the way, styled pyrophyllite by Kinahan in the Memoir and in his "Economic Geology," p. 59. Griffith, Haughton, and Scott (*op. cit.*, under Crohy Head) state that the Gartan soapstone passes into anthophyllite in the field. This steatite was again worked as soapstone in 1918 by Messrs. De la Hey Moores and W. Horner, of Londonderry.

Inniskil. 1″ 16. 6″ Donegal 51 S.E. An unworked band of steatite 4 feet wide occurs in a similar position on the edge of the granite mass in this townland, near Glendowan village, 1¼ mile S.W. of the head of Gartan Lough.

Crohy Head. 1″ 15. 6″ Donegal 56 N.E. Considerable beds of white and yellowish-white talc-schist occur among the Dalradian strata of Crohy Head, on the Atlantic coast W. of Dunglow. They have been quarried from the top of the cliff. Mem. 3, 4, etc., p. 46 (1891), suggests that the best part had been already removed; but a good deal of work has been done since then, and quarrying was carried on by the Orchard Refinery Co., of Belfast, in 1918.

R. Griffith, S. Haughton and R. H. Scott ("On the chem. and min. constitution of the granites of Donegal and of the rocks associated with them," Rep. Brit. Assoc., 1863, p. 59) give analyses of the Crohy Head talc crystals and of the massive soapstone. These are quoted in Mem. 3, 4, etc., p. 160, and the rock shows silica 60·24, ferric oxide 1·48, magnesia 35·14, and water 1·00. The water is low for an analysis adding up to 99·46; but there is no doubt that the material is talc. Water in the crystals is given as only 0·40 per cent. Cubes of pyrite militate against the quality of certain beds; but the quantity of steatite (talc-schist) in the cliff is greater than Kinahan and others have implied.

Claggan. 1" 73. 6" Mayo 65 S.W. This occurrence is in a little sea-inlet in the townland of Claggan, on Achill Island. It is noted on the 6" MS. map of the Geological Survey, and is referred to in Mem. 62 and 73, p. 22 (1879). The steatite occurs in bands and veins in a schistose series of rocks. Kinahan (p. 59) states that it was worked to some extent, that is, before 1889. Some work was again done about eight years ago.

Inishbofin. 1" 83. 6" Galway 9B S.W. and S.E., and 9D N.W. and N.E. (formerly Mayo 114 N.W. and N.E.). This island, off the Galway coast, is barely noticed in the Geological Survey Memoir, though its detailed geology is shown on the MS. 6" map. Kinahan, however (p. 60), calls attention to the working of a large mass of steatite here. Mr. W. B. Wright spent some days on the island in 1916, in response to enquiries that had reached the Geological Survey office, and he reports that the steatite belt is 20 to 100 yards wide, but is almost everywhere full of nodules of carbonates, from the size of a pea up to that of a man's head. Crystallised magnesite occurs among these nodules. The steatite band, as shown by Mr. Wright's revision of the map, runs from end to end of the island, and beneath the sea both west and east. It is regarded as of sedimentary origin, and shows no passage into the associated hornblendic igneous rocks. There are obvious difficulties in the way of working, including transport to the mainland.

Inishshark. 1" 83. 6" Galway 9C N.E., and 9D N.W. (formerly Mayo 114 N.W.). Kinahan (p. 60) compares the steatite here with that of Inishbofin. Mr. Wright (see above) reports that the steatite band of Inishshark is more free from carbonates, and is thus of better quality. It seems to be a continuation of that in Inishbofin.

CHAPTER XVI.

SULPHUR.

Iron Pyrites, commonly occurring in the form pyrite, which crystallises in the cubic system, has long been recognised as a "sulphur-ore," and as such has been largely raised in Ireland. Kinahan (p. 35) gives a long list of localities where this mineral occurs in noteworthy quantities, or where he believed it to occur under ferruginous gossan; this list should be consulted by prospectors, though only a few cases seem worthy to rank as mining propositions. The symbol FeS_2 is used on the Mineral Map at such spots as have been thought sufficiently important.

Ovoca (Avoca) Mineral Belt.

1" 130 and 139. Mem. 121, 130, p. 44 (1869), and 138 and 139 (1888). The abundance of iron pyrites in the copper mining belt of Co. Wicklow led to the development of the sulphur trade in Ireland in 1840. Hence this area may well be considered first. In earlier years (Weaver, 1819, p. 218) sulphur was extracted at the Cronebane mine by firing the copper pyrites in a closed retort, and collecting the sulphur in receivers. In 1840, however, as Smyth relates ("Mines of Wicklow," Rec. Royal School of Mines, vol. 1, pt. 3, p. 390, 1853), "the interruption of the sulphur trade with Sicily obliged the English manufacturers to turn their attention to the iron pyrites of the Ovoca mines." Kane, who is one of the first to record this development (1844, p. 213), says that the Government of Naples had placed an exorbitant price on sulphur. In May 1840 the Ovoca belt was meeting the crisis at the rate of 2,400 tons of ore a month, and 40,176 tons were exported to England for the manufacture of sulphuric acid in the year. Attempts to "economize" the ore on the spot in Wicklow did not prove successful. Smyth (p. 396) looked forward to further developments as "a bond so desirable and a boon so mutually useful" between England and Ireland, and he pointed out the importance of Wicklow "should our intercourse with Sicily be interrupted by war or other causes."

The Sicilian fiscal authorities (Kane 1845, p. 226) realised their error; but Smyth's table (*loc. cit.*, p. 391) of the amounts of iron pyrites shipped at Wicklow and Arklow from 1840 to 1852 shows that the total had risen at the end of that time to about 100,000 tons a year. In Min. Stat. 1860, p. xiv., we find that the total produce of iron pyrites in the United Kingdom had sunk to 52,000 tons in 1856; but revival was rapid between 1857 and 1869, 85 per cent. of the total being raised in Ireland (Rep. Controller Depart. Min. Resources, 1918, p. 22). In 1867 the production in Wicklow was 97,143 tons, 40,000 of which were supplied by the Ballymurtagh mine. In that year, however (Min. Stat. 1867, p. 50), Liverpool alone imported 25,273 tons from the Huelva mines in Southern Spain. In 1876, this figure had risen to 419,068 tons, and the effect may be seen in the reduction of the output from Ballymurtagh to only 945 tons in 1881. In the early years of the twentieth century the total Irish output had sunk to some 2,000 tons.

In 1883 a series of 23 plans and sections were placed by G. H. Kinahan in the Home Office (R 93) as of abandoned mines in the Ovoca district. The war of 1914-18, giving rise to the conditions forseen by Smyth sixty years before, called attention to the Ovoca district, and in 1919 and onwards the sulphur ore of Cronebane was raised by the Cronebane Mining Co. of Ovoca for use in the sulphuric acid trade in newly constructed furnaces at Arklow and Wicklow town.

The principal mines in the Ovoca valley that were concerned in the sulphur trade were **Connary** (often spelt *Connorree*, and other variants), **Cronebane, Tigroney** (often cited with Cronebane), **Ballygahan** and **Ballymurtagh.** Connary is in 1″ 130, and the other mines succeed one another down the valley from N. to S. in sheet 139. Ballygahan is in the townland of Ballygahan Lower. In some years in Min. Stat. a certain amount of the ore is separately quoted as "coppery." P. H. Argall ("Notes on the ancient and recent mining operations in the East Ovoca district," R.G.S.I., vol. 5, pp. 153 and 160, 1880) compares the iron pyrites lode in West Cronebane and Tigroney with that in East Cronebane and Connary, and notes that the coppery type of ore occurs in the former mines only. He gives (p. 164) the general sulphur content as 33 to 36 per cent. Gold, silver, nickel and cobalt occur in small quantities.

S. Haughton ("Notes on Irish Mines," G.S.D., vol. 5, p. 283, 1853) gives the sulphur-content of the iron pyrites of Ballymurtagh, whether coppery or not, as 35 per cent. On the older metallurgy of the pyrites, see Smyth, "Mines of Wicklow," p. 391. A MS. plan of the Tigroney sulphur ore lode by G. H. Kinahan is in the Geological Survey Office, Dublin.

Knockinoe, named in Min. Stat. 1866 as a Wicklow mine, raising 47 tons of iron pyrites, valued at £24, is probably the mine in **Knockanode** (6″ Wicklow 35 N.W.), just south of the bridge at the Meeting of the Waters. It lies outside the area of the Geological Survey special sheet of the Ovoca mines. Kinahan (p. 483) states that Knockanode was "extensively explored without success." Griffith (1861, p. 155) records that it produced lead and sulphur; the lead is on the authority of Weaver (1819, p. 219), who speaks of " slight strings."

The plan deposited in the Home Office series (R 284) is entered as that of a copper mine. It must not be confused with Knockanroe in Co. Tipperary (see below).

A mine raising iron pyrites and ochre is recorded as working in **Kilcashel** in 1917 (Home Office List of Mines). Sulphur ore and copper ore were raised in this townland prior to 1861 (Griffith, 1861, p. 154).

Knocknamohill Mines. 1″ 139. 6″ Wicklow 35 S.W. and 40 N.W. Kinahan (p. 34) says sulphur ore was worked in these. See under Iron.

I have been unable to trace the *Carrihans* mine stated in Min. Stat. 1880, to be in Wicklow, and to have produced 237 tons of iron pyrites. There is a townland of Carrahan in Co. Clare, 1″ 133; but the mine known as Carrahin near it seems to have raised galena and blende only. See under lead.

Clare Island. 1″ 73. Holdsworth ("Geology etc. of Ireland," p. 87, 1857) seems to be the only authority for the actual working of sulphur ore in Clare Island, which he says was proceeding when he wrote. The exhaustive Memoir on Clare Island, published by the Royal Irish Academy (p. 26, 1914), merely states that large pebbles of pyrite occur on the beach at Alnamarnagh, half a mile north of the emergence of the great mineralised fault-zone on the coast. One of these pebbles, collected by Mr. W. J. Lyons, was 8 cm. in diameter.

Gubnabinniaboy. 1″ 73. 6″ Mayo 75 N.E. In the townland of Bolinglanna, at the N.W. extremity of Clew Bay, where there is also an old copper mine, a lode of iron pyrites runs in from the shore, ¾ mile west of the houses of Bolinglanna that are shown on the 1″ Ordnance Survey map. The MS. 6″ Geological Survey sheet shows a shaft, and notes "mine (Sulphur pyrites) filled up." The lode is in the micaschist. Mem. 62, 73, p. 21 (1879), suggests that it was mined for arsenic as well as sulphur. The site is under the *o* of *Corraun* on the Mineral Map.

Derrylea. 1″ 93. 6″ Galway 36 N.W. In Mem. 93 and 94, p. 162 (1878), a large working is said to have been made at the east end of Derrylea lake, three miles east of Clifden, on a pyritous lode thought to be auriferous. The mine was unproductive. See also under Lead.

OUGHTERARD DISTRICT.

Mem. 93 and 94, pp. 160-165 (1878), Mem. 95, pp. 61-63 (1870), and Kin., p. 37, record a number of localities, north-west of Oughterard, where iron pyrites occurs in some quantity. Pyrrhotine is frequent. See also Glan Mines, under Copper. The symbol FeS$_2$ has been placed on the Mineral Map at a spot near the lake-shore east of Cornamona village (6" Galway 26 S.E.), where pyrrhotine and pyrite are associated (Mem. 95, p. 67).

A trial was made west of Doon Wood in Drumsnauv townland (6" Galway 39 N.W.; Mem. 93 and 94, p. 164). For the "Maumwee Lode" (6" Galway 38 N.E., Mem. 93 and 94, p. 163) see under Nickel. Shafts were sunk here, and the MS. 6" Geological Survey map notes that "Mining operations were carried on in the east of Teernakill South on pyritous ore and hornblende-schist." The shale is stated to contain "numerous irregular veins of pyrites."

Errisbeg. 1" 103. 6" Galway 63 N.W. This copper mine is recorded in Min. Stat. 1880 as raising, but not selling, 300 tons of coppery pyrites. The MS. 6" map records a trial shaft, and, on the coast, a quartz lode with coppery carbonate and iron pyrites.

Glanmore. This name in Min. Stat. probably refers to one of the "Mines of Glan," 1" 95 (see under Copper). Min. Stat. 1858, p. 67, records the raising of 800 tons of iron pyrites, valued at £700; but there is no return for 1859. There is a townland of Glenmore (6" Galway 90 N.E.) on the east side of Greatmans Bay, opposite Gorumna Island, where a great quartz reef is coloured in gold as a mineral lode on 1" 104. This is described in Mem. 104, p. 64, and does not seem to be pyritous. Glanmore appears in the Lists of Mines from 1865-8 as a mine of copper ore and pyrites (H. Hodgson).

Silvermines (Knockanroe). 1" 134. 6" Tipperary 26 S.E. The great wall of pyrite exposed here, close to the north border of Knockanroe townland, and south of the village of Silvermines, still presents an imposing spectacle; but no marked output has been secured. "Sulphur Mine" is engraved here on the 6" map. In M.C.I. Rep. 1840 ii, 1841 i, and 1842 i, the "sulphur ramp" is referred to, and the numerous excavations along it are cited as evidence of former mining activity in the district. In the Company's Report for 1843 i, the searches are said to have been concluded, and the machinery, which had been bought from the lessees, was removed to Knockmahon. Kane (1844, p. 189) describes the Company's efforts at *Knockeenroe*, and states that the current price of sulphur ore did not warrant further penetration of the mass in pursuit of the copper ore that might exist beyond. The M.C.I. was mostly concerned at Knockanroe with lead and copper. Weaver (1819, p. 243) had noted blende, and

Holdsworth ("Geology etc. of Ireland," p. 70, 1857) mentions zinc in connexion with the sulphur ore; but the development of the Silvermines zinc ores began with the discovery of the calamine deposits in 1858.

In 1858, 534 tons of iron pyrites were raised at Knockanroe, and 344 in 1859, valued at £170. There seem to be no later records.

A. B. Wynne ("Mining District of Silvermines," G.S.D., vol. 8, p. 249, 1860) describes the "Sulphur Mines" of Knockanroe, and gives a section across the lode (plate 15). See also Mem. 134, p. 38 (1861) under the name *Knockeenroe*. A new shaft was sunk here under the auspices of the Ministry of Munitions in 1917.

CHAPTER XVII.

ZINC.

Zinc was known in earlier times merely through its alloys, and it is possible that brass became discovered by the accidental substitution of zinc-blende for tinstone in the manufacture of bronze. Brass, however, was manufactured before the Christian era from copper and an earth fused with it, this earth being evidently one of the calamine ores (see art. ZINC, Encyc. Brit., ed. 11, vol. 28, p. 981). Queen Elizabeth granted a monopoly for the making of brass by the use of calamine. The metal zinc became known in Europe through a sample from India in 1597. Zinc-smelting was carried on in England in 1730, and at Liége in 1807. The blende that is so commonly associated with our lead ores then became for the first time a marketable commodity. The Report of the Controller of Mineral Resources for 1918, p. 18, notes that the total output of zinc ore from Ireland for the sixty years 1856-1915 was 17,865 tons, being twice that of Scotland, but less than one-twentieth of that raised in England.

Abbeytown. 1" 55. 6" Sligo 20 N.W. This ancient mine on Ballysadare Bay has already been described under Lead. The abundant association of blende with the deposit has led to its being worked for zinc also; it is mentioned as a lead and zinc mine by E. T. Hardman in 1880 (Proc. R. Dublin Soc., vol. 3, p. 12), and it was re-opened by Messrs. R. J. Kirwan and J. Mallagh of Sligo in 1917. Mr. H. J. Daly reported on Abbeytown to the Department for the development of Mineral Resources in that year.

Glengowla (*Glengola*). 1" 95. 6" Galway 54 S.W. This lode is described under Lead. A plan and a section of the mine occur in Mem. 95, p. 65 (1870). Some blende seems to have been raised.

Sheshodonnell. 1″ 123. 6″ Clare 10 S.W. This lode is marked with a zinc symbol on the 1″ geological sheet in latitude 53° 1′ and longitude 9° 5′ 40″ W. The name Sheshodonnell will be found on the 1″ Ordnance Survey sheet of 1900; the lode cuts the spur of the plateau north of the *d* of this word. F. J. Foot, in Mem. 114, 122 and 123, p. 27 (1863), says that a vein of "hydrated calamine" occurs here, associated with galena, cerussite, and fluorspar. It is 1 inch to 18 inches wide, and dies out abruptly north and south. Yellow or apple-green calamine (smithsonite) occurs, with variegated streaks in it. The locality was unknown to Griffith.

Connary (*Connorree, Connoree*). 1″ 130. 6″ Wicklow 35 N.W., and **Kilmacoo**, 1″ 130. 6″ Wicklow 35 N.E. Connary has raised copper, lead, and sulphur ores, and a good deal has been written on the "bluestone," a mixed grey ore that extends north-eastward into Kilmacoo (see Kinahan, p. 114). This deposit was noted by Weaver (1819, p. 215) as a mixture of galena, grey ore of antimony, blende, with "pyrites of copper, iron, and arsenic," and as not worth working, owing to the fact that no one mineral was predominant. It seems to be the zinc-bearing ore analysed by Jas. Apjohn in 1851 ("Upon the composition of a new variety of metallic ore, from the Vale of Ovoca," G.S.D., vol. 5, p. 134, 1853). He found FeS_2 24·97, FeS 7·33, PbS 19·13, and ZnS 46·62 per cent., and was inclined to regard the material as a definite compound. R. H. Scott ("On a new metallic ore from the Connorree Mines," G.S.D., vol. 8, p. 241, 1860), who comments on Apjohn's paper, gives the following analysis:—FeS_2 (iron pyrites) 50·653, FeS 12·338, ZnS 37·009 per cent., and regarded the two latter constituents as combined to form a ferriferous blende. Apjohn's specimen may have come from farther south; but the occurrence of lead sulphide was no doubt due merely to admixed galena. The "bluestone," however, became locally exalted to the dignity of a mineral species, under the name of "kilmacooite," and P. H. Argall ("Mining operations in the East Ovoca District," R.G.S.I., vol. 5, p. 164, 1880) quotes four analyses showing the proportions of the elements present. These dispose of any idea of constancy of composition. The zinc-contents are 8·16, 20·00, 24·00, and 49·08 per cent. The variable amounts of iron must be divided between pyrite, copper pyrites, and ferriferous blende. Traces of antimony and gold have been detected. Argall (*ibid.*, p. 160) is responsible for the introduction of "kilmacooite" into literature, though not for the invention of the term, and he fully recognised the composite nature of the "bluestone." Kinahan, in calling the associated rock "blue ground," raises a false comparison between it and the famous blue ground of South Africa. In 1884, C. R. C. Tichborne ("On an argentiferous galenitic blende at Ovoca," R.G.S.I., vol. 6, p. 296, 1886) made a careful examination of the Irish "bluestone," and from one analysis estimated the composition as ZnS 37·68, PbS 29·07, and Ag_2S 00·275 per cent. The silver was equal to 8·6 oz. to the ton, and the author lays stress on this as a source of profit. His paper is the best critical account of "kilmacooite," and includes a note (p. 298) by A. Ryder on the working of the material at Connary, at a time when it had been tested for about 120 fathoms eastward.

Min. Stat. record Connary as a zinc mine in 1877 (p. 172), worked by W. G. Du Bedat and Co., and give (p. 42) an output of 110 tons of ore, value £396. In 1878, 100 tons of ore were raised, and 184 in 1879. In 1885 the record is resumed with 98 tons from the bluestone, value £160; but it then ceases. A zinc record from "Wicklow" in Min. Stat. 1872 and 1873 (29 and 30 tons of ore respectively) probably refers to Connary.

Castletown. Lists of Mines 1862-5 give Castletown as a mine both of zinc and lead. The reference is probably to the northern mine of this name in 1" 123, 6" Clare 10 N.W. and S.W. See under Lead.

Ballyvergin. 1" 133. 6" Clare 26 S.E. Blende is said to have been raised here, as from other lead mines in the neighbourhood. See under Lead.

Carrahin (Carrahan). 1" 133. 6" Clare 34 N.E. This lead mine, in Lassana townland, is said to have raised blende. See under Lead.

Crowhill. 1" 133. 6" Clare 34 N.E. Blende is said to have been raised here with the galena. See under Lead.

Milltown. 1" 133. 6" Clare 35 N.W. Blende and galena are scattered here in the great calcite lode. The importance of the blende was pointed out in a report to the proprietors by Jas. Paull, who described the lode as 14 to 18 feet wide, with quartz, blende, limestone (presumably calcite), and argentiferous lead ore. At the part then reached, there were 4 tons of blende and 15 tons of galena per fathom, and it was held that the blende would pay expenses while further search for underlying galena was being made. I am indebted to Mr. E. R. Woakes, A.R.S.M., of Messrs. John Taylor and Sons, London, for lending the above report. Min. Stat. for 1858, record 14 tons 16 cwt. of zinc ore, value £164 10s.; this is the first mention of zinc from Ireland. In 1859 and 1860, 40 tons were raised each year. The record then ceases.

Silvermines District.

1" 134.   6" Tipperary 26 S.W.

After a long mining history (see under Copper, Lead, and Sulphur), the zinc ores in this district came into prominence. Weaver noted the blende in 1819 (p. 242), and Holdsworth ("Geology etc. of Ireland," p. 70, 1857) says that the Knockanroe lode consisted of ores of lead, zinc, sulphur, and copper. But the systematic raising of zinc ore followed on the discovery of electric calamine (hemimorphite) at the **Silvermines** mine south of the village of that name. A. B. Wynne (Mem. 134, p. 40, 1861) says that this mineral was detected by Capt. King in 1858. A specimen reached Dr. Jas. Apjohn in 1859, and was analysed and described by him as electric calamine in G.S.D., vol. 8, p. 157, in March of that year. The shaft from which it came was on the property of the General Mining Co. of Ireland, but was then abandoned. The pits whence calamine was subsequently raised are well shown in Wynne's map of the mining area (G.S.D., vol. 8, plate 15, 1860). By 1861, Wynne (Mem. 134, pp. 39 and 40) was able to describe the relation of the calamine ores, both smithsonite and hemimorphite, to the gossan of Silvermines, of which they form about 50 per cent. and to give a section showing the zinc deposit stretching southward as a replacement-bed in the magnesian limestone. J. B. Jukes ("On the way in which the calamine occurs at Silvermines," G.S.D., vol. 10, p. 11, 1864) emphasised this point, and suggested that the water spreading from the great fault decomposed the sulphide ore, forming soluble sulphates; these, in permeating the dolomitic limestone, gave rise to magnesium and calcium sulphates, which passed away in solution, and zinc and iron carbonates, which remained. The iron carbonate in time became oxidised to the ochre now prevalent in the gossan. "As most water contains silica, a certain amount of silicate of zinc has also been formed."

The records of output are all under the general name of Silvermines; but both this mine and **Gorteenadiha** appear as zinc mines under the Hibernian Development Co. in the List of Mines for 1906, both being said to be then discontinued. The ore raised was blende, smithsonite, and hemimorphite- the latter two being the calamine ores. Min. Stat. 1864, p. 43 (footnote), state that from 1859-61 the average produce of calamine was 100 tons, that very little was raised in 1862, and that only 47 tons were raised in 1863, when the total output of zinc ore from this mine is elsewhere stated as 3,892 tons. The product is sometimes recorded as zinc oxide, which was separated on the spot.

The first official record is 37 tons of calamine raised by

"the Mining Co. of Ireland" in 1859. The name should be "the General Mining Co. of Ireland." Five hundred and forty tons of ore were raised in 1860, and the output rapidly increased to 3,892 tons, mainly blende, in 1863, value £10,216. Zinc was then at the high price of £23 14s. per ton. Silvermines in that year produced one-third of the zinc output of the United Kingdom. In 1865, the ores are separately given as calamine 3,980 tons and blende 60 tons, a singular reversal of the conditions recorded for 1863; total value £8,140. In 1868, we have calamine 58 tons, zinc oxide 21 tons; total value only £246. In 1872, calamine 409 tons; zinc oxide, "manufactured from poor calamine ore," 196 tons. In 1874 the record closes with only 1 ton of ore, zinc being on an average down to £3 7s. per ton. In 1918, Mr. H. J. Daly reported on the Silvermines area to the Department for the development of Mineral Resources, and under his care three trial borings were put down. A lease from Lord Dunalley was subsequently obtained by the Silvermines Syndicate, who put down two additional bores on their own account. The sites of all these bores, and the location of the various old mineral workings in the district, are recorded on maps on the scale of 25" to one mile, prepared by Mr. T. Hallissy for the Geological Survey in 1917, and kept in the office.

Cloghatrida. 1" 153. 6" Limerick 20 S.W. See under Lead.

Shrelkald. I am unable to trace this name, given as that of an Irish mine with an output of 40 tons of zinc ore in Min. Stat. 1880, p. 45. It may be an accidental transference of a record from the Threlkeld zinc mine in Cumberland, which in the same year is returned as raising lead.

INDEX.

NOTE.—To assist reference from other works, older and even incorrect spellings of place-names are included in this Index, in addition to those adopted from the Maps of the Ordnance Survey.

A.

	PAGE
Abbey Island,	93
Abbeylands,	93
Abbeytown,	95, 143
Aghadown,	88
Aghamucky,	81
Aghatubrid,	88, 130
Aghnamullen,	100
Agnew's Hill,	70
Ailwee,	117
Aldfreck,	133
Allihies,	46
Annagh,	126
Annaglogh,	99
Annestown,	42, 127
Antimony,	14
Antrim Mines,	22
Araglin,	87
Ardclinis,	70
Ardmore,	127
Ardshins,	70
Ardtully,	43
Arigna,	77
Arklow,	31, 116
Audley Mines,	50, 53
Aughnagurgan,	100
Avoca,	30, 84, 139, 144
Avondale,	116

B.

	PAGE
Ballard,	82, 130
Ballinafinchogue,	113
Ballinafunshoge,	111, 113
Ballinagoneen,	114
Ballingarane,	124
Ballinglanna,	124
Balliness,	88
Ballinoe,	37, 122
Ballintemple,	115
Ballintoy,	23
Ballintreddedy,	123
Ballintredida,	15, 123
Ballinvally,	63, 65
Ballinvirick,	15, 123

	PAGE
Ballonakill,	77
Ballybig,	133
Ballyboley,	70
Ballyboy (Ballybay),	97
Ballycapple,	82
Ballycastle area, iron,	67
Ballycohen,	39
Ballycomisk,	50
Ballycoog,	86
Ballycormick,	86
Ballycorus,	107
Ballycouge,	86
Ballycullen,	114
Ballycummish,	50
Ballycummisk,	50
Ballydahab,	51
Ballydanab,	51
Ballydehob,	51, 128
Ballyedmond,	133
Ballyedward,	133
Ballygahan,	35, 84, 140
Ballygowan South,	122, 123
Ballygown,	123
Ballyhickey,	119
Ballyhighland,	125
Ballyhourigan,	37
Ballyhurly,	121
Ballylagan,	71
Ballylaggin,	71
Ballylig,	133
Ballylumford,	73
Ballymacarbry,	39
Ballymagrorty,	93
Ballymalone,	81
Ballymaquiff,	116
Ballymartin,	71
Ballymoneen,	36, 85
Ballymurtagh,	35, 84, 140
Ballynagowen,	114
Ballynakill,	39, 77
Ballynasissala,	42
Ballyness,	88
Ballynoe,	37, 122, 123
Ballypalady,	71
Ballysadare,	95

150 INDEX.

	PAGE		PAGE
Ballyshannon Mines,	93	Caminches Mine,	47
Ballysteen,	123	Canrawer West,	103
Ballyvaughan Mines,	117	Cappagh,	53
Ballyvergin,	36, 117, 145	Cappaghglass,	53
Balteen,	55	Carahan,	118
Bannishall,	52	Carberry West,	52
Banistown,	125	Cargan,	24, 71
Bantry district,	17	Carhoon,	116
Bantry Lead Mine,	128	Carndaisy,	75
Barard,	71	Carnlough,	71
Baravore,	17, 114	Carrahan, Carrahin,	118, 145
Baronstown,	104	Carravilleen,	53, 58
Barrastown,	125	Carrickagarvan,	100
Barravone, Barravore,	17, 114	Carrickartagh,	16
Barretstown,	125	Carrickfergus area,	134
Barrystown,	125	Carrickmacross area,	65
Barytes,	15	Carrickmore,	67
Bauxite,	22	Carricknahorna,	93
Bawnishall,	52	Carrigacat,	55
Bay Mines,	71, 74	Carrigcrohane,	44
Bealkelly,	81	Carrignahelty,	79
Bearhaven Mines,	45	Carrihans,	141
Beauparc,	28	Carrowgarow,	102
Behaghane,	44	Carrowgarriff,	102
Belderg,	26, 27	Carrowmore,	91
Belfast,	65	Carrowtrasna,	137
Belleek district,	62, 76	Carryhugh,	98
Benderg,	27	Carysfort Mines,	31, 115
Berehaven Mines,	45	Cashla Bay,	104
Bluestone,	33, 144	Castleblayney,	100
Boate, Gerard,	9	Castle Caldwell,	76, 131
Bog with Copper Ore,	60	Castle Freke,	60
Bog Iron Ore,	88	Castlegrove,	92
Bog Mine,	53	Castleknock,	30, 105
Boleagh,	52	Castlepoint,	60
Bolinglanna,	27, 79	Castleshane,	14
Bond,	97	Castletown,	117, 120, 145
Bonmahon,	41	Castleward,	94
Boulysallagh,	52, 129	Church Mine,	105
Brandon Barytes,	20	Claggan,	138
Brockagh,	110	Claragh,	79
Broughshane,	71	Clare county, lead,	116
Brow Head,	52	Clare Island,	141
Brownstown,	28	Claremount,	29, 103
Bunmahon,	41	Clasagh,	121
Burleigh Hill,	136	Clay,	98
Burnt Irish,	60	Clea,	98
Burren Mines,	117	Cleengort,	61
		Cleenrah,	80
C.		Cleggan,	29
		Clegnagh,	23
Caherglassan,	116	Clements Mine,	102
Caherglassaun, Cahenglassaur,	116	Cloan,	46, 47
Cahermeeleboe,	48	Cloghane,	59
Cahermore,	128	Cloghatrida,	124, 147
Caher West,	127	Cloghcor,	71
Caim,	125	Cloghleagh,	80, 130
Callaros,	52	Cloghran,	105
Calliagh,	78, 129	Cloncurry,	61
Camaderry,	110	Clonetrace,	71

INDEX

	PAGE		PAGE
Clonkeen,	114	Cusackstown,	28
Clonligg,	94		
Clonmines,	126	**D.**	
Clontarf,	105		
Clontibret,	14, 96	Dalkey,	106
Clontua,	43, 127	Deehommed,	76
Cloontoo,	43, 127	Deel Mountain,	79
Cloosh, Clooshgereen,	17, 103	Derreengreanagh,	19, 55
Cluin,	47	Derreennalomane,	20, 55
Coad Mines,	44	Derreennatra,	55
Coagh,	66	Derrycarhoon,	55
Coal Island,	76	Derrycarne,	80
College Mine,	98	Derryganagh,	18
Comaderry,	110	Derryginagh,	18
Comnanmoor lode,	47	Derrygranagh,	18
Coney Island,	53, 95	Derrygreenagh,	18
Conham,	61	Derrygrenach,	18
Conlig,	94	Derrylea,	102, 141
Connaree,	33	Derryloaghan,	61
Connary,	14, 33, 84, 115, 140, 144	Derrynea,	104
Connery,	33	Derrynoose,	98
Connoree,	33, 84, 115, 144	Dhurode,	55
Connorree,	140, 144	Doagh,	71
Conny Island,	95	Dolphin's Barn,	105
Cookstown area,	74	Doneen,	21, 46, 60, 129
Coolartragh,	97	Donegal county, bog iron,	89
Cooleen,	122	Dooey Point,	61
Coolruntha,	37	Doolough,	62
Coom Mine,	47	Dooneen,	46
Coony,	53	Dooros,	29
Cooragurteen,	52	Down county, lead,	94
Coosheen,	54, 88	Downshire Mine,	136
Copper,	26	Dreenalamane,	20
Copper ore in bog,	60	Drumgill,	66
Coreen,	71	Drummeland,	98
Cornalough,	100	Drumreen,	92
Cornamucklagh South,	96	Drumshambo,	78
Corraun Mines,	27	Drumslig,	87
Correen,	25, 71	Dublin district, lead,	104
Cosheen,	54	Dublin Society,	9, 10
Coveen,	71	Dunbeacon,	20
Craigahulliar,	71	Duncrue,	134
Creevelea,	77	Dundalk,	100
Cregg,	17, 103	Dundrum,	95
Creggan,	101	Duneany,	72
Croaghan,	99	Duneen,	21, 60, 129
Croghan Kinshelagh,	63	Duneen Bay,	21, 60, 129
Crohy Head,	137	Dungannon,	76
Crommelin,	72	Dungonnell,	72
Cronebane,	33, 84, 140	Dunluce,	72
Crookhaven,	54	Durrus,	20
Crossreagh,	72	Dysart,	82
Crowhill,	118, 145	Dysert,	127
Crow Island,	40		
Cuilcagh,	77	**E.**	
Cullaleen,	72	Eden,	136
Cullenagh Hill,	81, 82	Eighterross,	92
Cullentragh Park,	114	Elgenany,	72
Cullinane,	25	Elginny,	25, 72
Curraghduff,	29	Enaghan,	80

INDEX

	PAGE		PAGE
Errisbeg,	30	Gortavallig,	56, 58
Errisberg,	30	Gorteenadiha,	36, 123, 146
Essathohan,	24, 72	Gortinee,	79
Evishacrow,	24, 72	Gortnadine,	36, 123
Evishnablay,	72, 73	Gortnageeragh,	72, 73
Evishnacrow,	72	Gortshaneroe,	37, 122
		Grallagh,	87

F.

		Great Cappagh,	53
Feakle,	81, 89, 130	Griffith, R. J.,	9, 11
Felspar,	61	Gubnabinniaboy,	141
Fingal,	61	Gurteenadya,	36, 123
Fintown,	92	Gurtnadyne,	36, 123
Flood Hall,	124	Gurtyvallig,	56
Foxrock Mine,	110	Gypsum,	65
French Park,	135		

H.

		Hawk Rock,	110
		Hibernian Mine,	129

G.

Galena as a silver ore,	90	Holdsworth, J.,	10
Galway county, bog iron,	89	Hollyford,	39
Galway Mines,	104	Hollyhill,	56
Garrough,	44	Hollyrock lode,	111, 112
Garryard East,	122	Hollywood,	114
Garryard West,	122	Holyford,	39
Garrykennedy,	120	Hope Mines,	100
Gartan,	137	Horse Island,	26, 56
Geevraun,	26	Hunt, R.,	12
Geological Survey,	10		
Glan,	29		
Glanalin,	56, 58	### I.	
Glandore Mines,	60, 88, 130	Ihron,	48
Glanmore,	29, 129, 142	Imail,	82
Glann Mines,	29	Inishbofin,	138
Glaslough,	77	Inishdooey,	132
Glebe,	72	Inishshark,	138
Glenaboghil,	92	Inniskil,	137
Glenanlin,	56	Irhin,	48
Glenariff,	71, 72	Irish Consols,	61
Glenarm,	25, 72	Irish Hill,	25, 73
Glencar,	16	Iron,	66
Glencarberry,	16	Island Magee,	73
Glencarbury,	16	Islandmore,	73
Glendalough,	84, 110, 112		
Glendasan,	110	### K.	
Glendree,	117		
Glengola,	102, 143	Kamminches,	47
Glengowla,	102, 143	Kane, R.,	10
Gleniff,	16	Keallogue,	47
Glenmalur,	113	Kealoge,	47
Glen Mine,	41	Keeldrum,	91
Glenravel Mines,	72	Kilbarry,	52, 55
Glenties,	93	Kilbrain,	93
Glentogher,	91	Kilbreckan,	15, 119
Glentubert,	14, 96	Kilbricken,	15, 119
Glin,	86	Kilbrickenite,	15
Gobb,	68	Kilbride,	80, 82, 130
Gold,	63	Kilcashel,	34, 85, 141
Goldmines,	37	Kilcoe,	52, 57
Gold Mines Valley,	63	Kilcolman,	86
Gortacloona,	128	Kilcrohane,	58
		Kildavin,	80

INDEX

	PAGE		PAGE
Kildrum,	. 91	Lisduff,	. 129
Kilduane,	. 41	Lisglassan,	14, 96
Kilkilleen,	. 59	Lisglassin,	. 14
Kilkinnikin West,	. 128	Lissan,	. 74
Killeen,	38, 58	Lochstuckagh,	. 101
Killeen North,	. 58	Loghill,	. 86
Killelton,	. 43	Long Island,	. 58
Killiane,	. 17	Lough Allen coalfield, iron,	. 77
Killiney Hill,	. 106	Lough Anure,	. 132
Killoveenoge,	. 128	Lough Corrib,	. 102
Killovenogue,	. 128	Lough Lara,	. 132
Killowen,	. 127	Loughnambra'dan,	. 92
Killygreen,	24, 73	Lough Neagh ironstone,	. 70
Kilmacoo,	33, 144	Loughshinny,	. 30
Kilmacooite,	33, 144	Lough Unshin,	. 62
Kilmainham,	. 105	Luganure,	. 110
Kilmocapogue,	. 18		
Kilmore,	120, 121	**M.**	
Kilmorie,	. 120	Magheramenagh,	. 76
Kilmurrin,	. 42	Magheramorne,	. 133
Kilmurry,	. 120	Magpie Mine,	. 33
Kilrean,	. 93	Mahoonagh,	. 124
Kilwaughter,	. 73	Maiden Mount,	. 135
Kinahan, G. H.,	. 12	Main,	73, 124
Kingscourt area, gypsum,	. 65	Mains,	. 73
Knockadrina,	. 124	Malagow,	. 61
Knockanode,	34, 115, 140	Mallavoge,	. 52
Knockanroe,	122, 142	Mallinmore,	. 79
Knockaphreagaun,	. 118	Manganese,	. 129
Knockatillane,	80, 130	Maumwee lode,	132, 142
Knockatrellane,	. 39	Mayne,	. 124
Knockboy,	. 73	Meanus,	. 126
Knockmahon,	. 41	Milltown,	118, 145
Knockmahon area, copper,	. 41	Mineral Map, principles of,	. 6
Knockmalaur,	. 61	Mineral Statistics,	. 12
Knocknamohill,	36, 85, 141	Mining Company of Ireland,	. 13
Knocknacran,	. 65	Mining Record Office,	. 11
Knockybrin,	. 92	Mizen Head,	. 59
		Mogouhy,	. 117
L.		Moll Doyle Mine,	. 110
Lackamore,	. 38	Molybdenum,	. 131
Lackavaun,	. 55	Molyneux, T.,	. 9
Lady's Cove,	. 43	Monachoe,	. 42
Laheratanvally,	. 59	Monanoe,	. 119
Landore,	. 60	Moneyhoe,	. 41
Lanmore,	. 129	Moneyteige,	36, 86
Lansdown Mine,	. 129	Mountain Mine,	22, 46
Larkhill,	. 62	Mountcashel,	72, 73
Lassana,	118, 145	Mount Gabriel,	21, 59
Laterite ores,	. 69	Mount Mapas,	. 106
Lead,	. 89	Mount Peru,	. 107
Leggs,	. 76	Mountrath,	81, 82
Leighcloon,	. 59	Moyriesk,	. 120
Leinster coalfield, iron,	. 81	Muckross,	. 40
Lemgare,	. 99	Muckruss Head,	21, 60, 129
Lemnagh More,	. 23	Mullantiboyle,	. 93
Lemonfield,	. 103	Mullavoge,	. 52
Letter,	. 21	Mullenmore North,	. 79
Libbert Mine,	. 25	Mullentybogh,	. 93
Lisdrumgormly,	. 99	Murvey,	. 132

INDEX

N.

	PAGE
Nanmor,	61
New Hero Mine,	111
New Keallogue,	47
New Mine,	136
Newtownards,	94
Newtown Crommelin,	73
Nickel,	132
North Hero Mine,	111
North Ruplagh,	110
North Tankardstown,	42

O.

	PAGE
Old Hero Mine,	110
Old Luganure,	110
Oolahills,	39, 124
Oola Mines,	39, 124
Oranmore,	104
Orblereigh,	74
Ordnance Survey Maps,	10
Oughterard,	103
Oughterard district—	
Lead,	102
Sulphur,	141
Ovoca Mines—	
Copper,	30
Iron,	84
Sulphur,	139
Zinc,	144

P.

	PAGE
Pallaskenry,	37
Parkmore,	73
Pollboy,	26

Q.

	PAGE
Quin Mines,	117

R.

	PAGE
Raspe, R. E.,	40, 126
Rathkenny,	73
Rathlin Island,	73
Rathnaveoge,	37
Red Bay,	74
Redhall,	133
Redhills,	79
Red Rock,	74
Renville,	104
Ringabella,	128
Rinville,	104
Rinvyle,	104
Roaring Water,	59
Rock Lodge,	86
Rock-salt,	133
Rossbeg,	76
Ross Island,	39
Rooska East,	128
Rostellan,	87

	PAGE
Royal Dublin Society,	9, 10
Ruplagh,	110

S.

	PAGE
Salterstown,	28, 101
Scart,	19, 55
Schull Bay,	54
Shallee,	122
Shanagarry,	127
Shanagolden,	86
Shane's Hill,	74
Shankhill,	108
Shankill,	108
Sheeffry,	101
Sheffry,	101
Sheshodonnell,	144
Shionagree,	61
Shrelkald,	147
Silver, see under Lead Mines,	89
Silver in Lead ore,	90
Silvermines district—	
Barytes,	17
Copper,	36
Lead,	121
Sulphur,	142
Zinc,	146
Silvermines Mine,	123, 146
Skeagh,	21
Skeaghanore,	52
Skull,	59
Skull Bay,	54
Slieve Anierin,	77
Slieve Gallion,	75
Slievenanee,	74
Smyth, W. W.,	10
Solomon's Drift,	25, 74
Spanish Cove,	52, 55
Srahmore,	27
Sralaghy,	27
Statistical Surveys,	10
Steatite,	137
Stewart, Donald,	9
Straid,	25, 73
Strangford,	94
Sulphur,	139
Sutton,	129
Swanlinbar,	77

T.

	PAGE
Tamlat,	100
Tankardstown,	42
Tassan,	97
Tattyreagh,	100
Tawnycrower,	101
Teernakill South,	29, 132, 142
Tennant Mine,	136
Termoncarragh,	62
Ternakill,	29
Tiernakill,	132
Tigroney,	34, 84, 140

INDEX

	PAGE
Tirgan	75
Tomgraney,	81
Tonagh,	97
Tormore,	16
Tragh-na-mban,	48
Trostan,	74
Truska,	29
Tuftarney,	25, 74
Tulla Mines,	117
Tullybuck,	14, 96
Tullycall,	74
Tullydonnell,	28
Tullynawood,	100
Tullyratty,	95
Twigspark,	27, 93
Tynagh,	116
Tyrone coalfield, iron,	76

U.

Unagh,	74
Urbalreagh,	24, 74
Urblereigh,	74
Urhin,	48

V.

	PAGE
Valencia,	43
Van Diemen's Land Mine,	110

W.

Walterstown,	28
Weaver, T.,	10
West Cahir,	127
West Carbery district,	48
West Cronebane,	33
West Luganure,	110
Westport,	129
Wheatfield,	105
Whitespots,	94
Wicklow,	116
Wicklow Hills, lead,	109
Woodford,	81, 89
Woodstown,	43

Z

Zinc,	143

Wt. T.2150—S.98. 600. 10. 21. B. & N., Ltd.—Group 4